MW00860943

ENCHANTED BY DAPHNE

FRONTISPIECE. In my office at the University of Michigan, September 1983.

Enchanted by Daphne

THE LIFE OF AN EVOLUTIONARY NATURALIST

PETER R. GRANT

PRINCETON UNIVERSITY PRESS

PRINCETON & OXFORD

Published by Princeton University Press
41 William Street, Princeton, New Jersey 08540
99 Banbury Road, Oxford OX2 6JX

press.princeton.edu

All Rights Reserved

ISBN 9780691246246
ISBN (e-book) 9780691246291

British Library Cataloging-in-Publication Data is available

Editorial: Alison Kalett and Hallie Schaeffer
Production Editorial: Theresa Liu
Jacket/Cover Design: Wanda España
Production: Danielle Amatucci
Publicity: Matthew Taylor and Kate Farquhar-Thomson
Copyeditor: Amy K. Hughes

Jacket image: Courtesy of B. R. Grant.

This book has been composed in Classic Arno

Printed on acid-free paper. ∞

Printed in the United States of America

10 9 8 7 6 5 4 3 2 1

For Rosemary
The perfect companion on my journey in adult life

CONTENTS

PREFACE

THE PERSON who inherits a career from a parent walks a straight line, knowing the future in light of the past. My path, in contrast and like those of most people, has been crooked and erratic, reoriented at crucial points by idiosyncratic factors and completely unpredictable. I chose my path, and my path chose where to take me. That is one reason I find my own life so interesting and am writing about it. A second reason is more prosaic—it is to enjoy remembering, recollecting, and resurrecting memories, and giving them coherence by narrating them, putting the pieces together and making them a story by arranging them linearly. By doing so I describe the life of a scientist, a biologist who works in the field studying ecology and evolution with his biologist wife and spends most of his life in society. It is a story quite unlike any other I have read.

Autobiographical stories are well-dressed portraits subtly or plainly edited, revealing self in the best possible light. When I began to write my own, I thought I could avoid narcissism by writing in the spirit of "I am a camera," the device used by Christopher Isherwood in *Goodbye to Berlin* (Isherwood 1939) to report what the eye sees and the brain records in an attempt at complete objectivity. The attempt was as futile as it was naive, as life stories are told to listeners and readers who want the narrator to be more than a camera. I became aware of a second contrivance, that writing about self is in fact a schizophrenic act, since I the writer am also I the actor. Even if the writer is not particularly impressed by the actor, he is occasionally embarrassed by him, so the camera fails us, the images become too blurred to be recognizable and so are deleted or modified. This seems a good metaphor for my memory.

I take it for granted that memory is unreliable, selective, and at times likely to be completely wrong. I could have called my story *Unreliable*

Memoirs, but that has been taken (James 1980). I am also aware of being prone to the sixth of seven sins of memory, that of bias, a sin of misremembering the past so as to make it more consistent with current knowledge and beliefs (Schacter 2001). My mother was similarly biased by an inventive memory. As Alice Kaplan writes, "It is helpful to remember that memory is a novelist" (Kaplan 2012, 174). However, I do have two safeguards against falsely created memory. One is the superior memory of Rosemary, my wife, and the second is a set of diaries from boyhood (1947–57) together with an electronic diary I have kept since the end of 1991.

This book has its origin in Covid-19-induced lockdown. Soon after SARS coronavirus-2 started to spread and cause havoc around the world, I read, or rather reread, Albert Camus's *The Plague* (1960), the most vivid account of a lockdown I have ever read, although, or perhaps because, it is written as a fable (O'Brien 1970). Having enjoyed the modern relevance and much else besides, I turned once again to Daniel Defoe's eighteenth-century novel *A Journal of the Plague Year* (1966 [1722]). Defoe's journalistic style, and perhaps Camus's as well, surely influenced the way I approached the question of how best to tell the story of my life, because when I started to write in September 2020, the two accounts were still uppermost in my mind. Beyond the temporal framework, there is no connection with either author except for curious coincidences that begin with a pirate, William Dampier.

Dampier influenced Daniel Defoe in both writing style and subject matter (Eiseley 1970). Dampier's travels, exploits, and geographical and occasional biological observations are well known because he kept a journal, preserved it in bamboo sealed at the ends with wax and, on returning to England, wrote a book, *A New Voyage Around the World*, "composed of a mixt Relation of Places and Actions, In the same order of time in which they occurred" (Dampier 1927 [1697], preface). This remarkable buccaneer-explorer, a keen observer and loner among a crowd of ruffians, impressed the diarist John Evelyn ("He seemed a more modest man than one would imagine by relation of the crew he had assorted with" [Hasty 2011, 40]) as well as the poet Samuel T. Coleridge ("a rough sailor, but a man of exquisite mind" [Hasty 2011, 54]).

As he records in his book, Dampier was one of the pirates who used the Galápagos—the Enchanted Isles—as a temporary hideout in 1684 after raiding Spanish towns and ships along the west coast of South America. Two years later, and in the company of a different set of pirates, he and his colleagues retreated to the Islas Marías, often called the Tres Marías (Three Marys), off the west coast of Mexico, to lick their wounds after marauding skirmishes on the mainland, some successful, others disastrous. It happens that I did PhD research on the Tres Marías and spent many years doing research on the Galápagos. The coincidence with Dampier is remarkable because I know of no other person who has camped in both places. Coincidentally, at about the same time that Dampier visited the Galápagos, the first house (dated 1683) was built in the municipality of Princeton, where I live. I feel strangely connected through a web of influences and coincidences to this piece of history.

My story, the theme of this book, is a journey through time of a whole life, and in the telling I am inviting the reader to be a companion. The readers I have in mind are scientists who know my work, young aspiring scientists, and nonscientists who might be curious about the background, development, and life of a scientist. In adult life, the journey took not one path but two and switched from one to the other. For many years Rosemary, our children, Nicola and Thalia, and I visited Galápagos every year to carry out studies of Darwin's finches. Life for most of the year in society contrasts so much with our mainly solitary life in the Galápagos that they constitute two pathways. I have kept them largely separate in different chapters, different segments of bamboo.

A few words of explanation about the title. Daphne plays a central role in my life. In Galápagos there are two islands called Daphne, Daphne Major (Mayor) (Fig. P.1) and Daphne Minor (Chica). Neither has ever been inhabited. I climbed Daphne Minor once with the help of mountaineering friends and worked on Daphne Major every year for forty years. Daphne Major was named after the British naval vessel HMS *Daphne*, which sailed in Galápagos waters in 1846 (Woram 1989; Grant and Grant 2014), and Daphne Minor was named by the naturalist William Beebe (1924). In Greco-Roman mythology, the young naiad Daphne narrowly escaped being "kissed" by Apollo, when Zeus,

FIGURE P.1. **Upper**: Daphne Major Island, Galápagos. **Lower**: Daphne crater.

responding to a plea for help, changed her into a laurel tree. Subsequently, a wreath made of laurel leaves was awarded as a prize at the Pythian Games, held every four years at Delphi in honor of Apollo but equally in memory of Daphne. Hence there is a remote connection between Daphne, the Delphi prizes, and the laurel wreaths placed on Rosemary's head and mine when we received PhD degrees at Uppsala University in 1986 for accomplishments on Daphne Island. I like to think I first heard of Daphne much earlier, in 1938, when I was less than two years old and listened to a gramophone record of the jazz guitarist Django Reinhardt playing a piece of music he wrote called "Daphne." A siren for the future!

1

Daphne

November 23, 1905 Sailed for Daphne Island in the morning . . . the
mate lost the skiff so we had to return to the ship and sail after it; we got
everything straightened out about 1:30 after capsizing the two other
boats in a series of maneuvers and landed again at about 2 o'clock.

*RACE WITH EXTINCTION: HERPETOLOGICAL FIELD
NOTES OF J. R. SLEVIN'S JOURNEY TO THE GALÁPAGOS,
1905–1906 (FRITTS AND FRITTS 1982)*

FEBRUARY 5, 1998 There is no anchorage; the volcanic rocks slope
steeply into the sea and rapidly disappear (Fig. 1.1). Nor is there a beach.
We land on a barnacle-covered wave-cut platform, very roughly flat and
measuring little more than one yard by two. A volunteer helper stands
to receive goods from a crew member of the *Pirata* at the front of a
panga (Zodiac) as it rises with the surf and twists with the current. Another volunteer, a marine biology student from Guayaquil, Ecuador, is
halfway up a pockmarked, eight-foot-tall vertical cliff, acting as interme-
diary between his companion at the landing (or me) and Rosemary at
the top. It's a crucial position in the chain; the person has to be strong,
fit, and flexible. Men are not the only ones to have occupied the position, but Rosemary and I are now too old to do the whole exercise safely
by ourselves; that is why we have assistants.

FIGURE 1.1. Daphne Major landing. **Upper left**: The wave-cut, barnacle-covered platform at low tide; arrows indicate "step." **Lower left**: Peter Boag and Laurene Ratcliffe leaving Daphne, 1976. **Right**: Departure when sea is calm (photo M. Wikelski). (Upper left and right from Grant and Grant 2014, Fig. 1.3.)

Seabirds—gulls, boobies, and tropicbirds—fill the air with their whiteness and cries. There is movement everywhere. I am perched on an adjacent rock to help, as an additional receiver, and all goes well until I drop my guard, and the swell gently lifts me up and off my perch without any intention of putting me back there again. It is, after all, an El Niño year, and a surf several feet high is to be expected. It is the first time this has happened to me and is destined to be the only time.

By the time we have completed the unloading, there will be thirty or forty green plastic five-gallon bottles of water (*chimbuzos*) at the top of the cliff, to be shifted to shady shelters and cavelets for protection against the sun. The arid little island gives us neither food nor water. The *chimbuzos* will be accompanied by six to ten white metal waterproof boxes with our food for a month: rice, cans of tuna, packets of dried soup, cookies, sugar, coffee, and powdered milk that is designed for in-fants elsewhere. A few loaves of bread, a stick of bananas, and a few potatoes will be the luxury fresh foods for the first few days. All will need to be moved to caves for shelter from the sun.

We put up a shade over an open cave that faces the sea and the island from which we have come, Santa Cruz. This will be the kitchen. The tent—the bedroom—will go up later on a small patch of semi-flat ground next to a large spread of prickly pear cactus perched at the top of a steep slope to the sea. We are excited to be back for another field season, our twenty-sixth, to study Darwin's finches that we know so well, but we are also very busy setting up camp under the equatorial sun, and I have no time to do anything other than register the fact that *Geospiza scandens* 10105, a Common Cactus Finch in his fifteenth year, is singing on his cactus bush, as usual.

2

Childhood

A KALEIDOSCOPE OF MEMORIES

I ENTERED the world in Upper Norwood, a suburb of South London where all the males in my lineage had lived going back to 1828 or earlier. It is a suburb created out of the Northwood, an extensive woodland that was home to charcoal burners, smugglers, and Gypsies* in earlier times. In fact, "It has been said that if you look into the eyes of a Norwood local you look into the eyes of a gypsy" (Warwick 1972, 28). I don't know about that—my genome might be revealing one day—but I do know my life has been nomadic.

On November 30, 1936, exactly five weeks after I was born, the Crystal Palace burned down. My auntie Vi told me I was held up to the window in our house (47 Harold Road) on Beulah Hill to see the flames, which one witness proclaimed reached three hundred feet high (Warwick 1972, 241). I wish I could say I remember it and the palace itself—an exhibition hall made of glass that celebrated the achievements of Victorian England when built in 1851.

Memories of my early life are meager. I believe my earliest memory is of smelling marigolds in Park Court in Sydenham, where my mother lived after my parents' separation. I remember remembering the heavy, distinctive aroma on a postwar visit, but when was the first occasion?

---

* "Gypsy," used at the time as a descriptive term, is now considered perjorative, and "Roma" is generally preferred.

Having rehearsed and revisited those early memories many times, I fear they are prone to error that is repeated, like a cultural mutation. Another undated but early memory was putting the arm of a record player gently down on a slowly turning 78 rpm record on the turntable of an HMV (His Master's Voice) gramophone. I remember doing this because I was frequently reminded by my father and grandmother how good I was at it when I was only two or three. Jazz music was my father's passion. Bix Beiderbecke played the cornet on several records, Stéphane Grappelli the violin, and Django Reinhardt the guitar.

My early life was dominated by two major events. First, my mother and father (Fig. 2.1) separated in late 1938 or early in 1939 and subsequently divorced. I was awarded by the court to my father, who was the "wronged party." Second, World War II began in September 1939, and from age three to four I lived with my cousin Brian's family, the Jackmans, at 34 Briarwood Road in Stoneleigh, near Epsom, in Surrey. Perhaps the separation from my parents is the reason I don't remember much. Another problem is that after the war I frequently stayed with the Jackmans, and the two sets of memories have coalesced into one or become so intermingled that I don't trust my placement of many of them. But one anecdote I can assign to 1940 is the story of how I, at age three, absconded with Janet, the two-year-old sister of Brian's friend Geoff Lanegan. We pushed her doll's pram up the road and onward, until we were found by a friendly policeman and taken to a police station to find out where we lived, and thence home.

Life with the Jackmans was very enjoyable because I had a playmate in Brian, and we had the freedom to explore to a degree that is probably denied almost all young children in today's more dangerous world, except in rural or remote areas. We principally explored Nonsuch Park, once King Henry VIII's hunting grounds, climbed trees, collected birds' eggs, and caught Common and Crested Newts with a wriggling worm at the end of a line from a makeshift fishing rod. These we put into a jam jar filled with water so that we could admire their colors and dragon-like crests. At one time Brian had read about Gypsies eating earthworms, so of course we had to try it and convince ourselves they were very good. They are actually rubbery. Inadvertently, we were helping

FIGURE 2.1. **Upper left**: With my father at Margate, UK, 1937. **Lower left**: With my mother at my grandparents' house, 319 Blandford Road, Elmer's End, Kent, UK, September 1938. **Right**: At same place and time.

to prime our respective immune systems for the future, not to mention challenging our microbiomes. We became very proficient in finding caterpillars of the large and colorful hawkmoths by cueing in to their fecal pellets on the sidewalk. Some were named after the plants they fed on: privet, lime, poplar. We kept them in boxes, fed them their respective leaves, recorded when they changed into pupae, and then marveled at the colorful and very different adults that emerged.

Auntie Lil was a kindly mother figure, never more so than when she treated a wasp sting on my scrotum with Reckitt's Blue to kill the pain; incidentally, the sting generated another priming of the immune system! In contrast, Uncle Stan was a stern disciplinarian who displayed little affection to anyone. Some of these memories belonged to my fourth year, but our vermivory probably came later. At the end of that year, Uncle Stan declared he could afford to keep me no longer. I remember the tension this caused in the family. The meeting with my father took place in the mysterious living room, which was otherwise permanently locked and preserved for special occasions. The upshot was, I left.

Like many children in urban and suburban Britain, Brian and I were evacuated during the war to rural areas. The immediate stimulus was the Blitz, a daily bombardment of London by German planes that began toward the end of 1940 and went on for eight months. Evacuation was the government's policy to protect us, the leaders of the next generation (Wicks 1988). Brian was sent to Cornwall, where he had a miserable time with a penny-pinching farmer's family and was worked to the bone (Jackman 2021). He did not go to school. He felt abandoned by his parents, a feeling forcefully echoed by the writer Oliver Sacks under much worse circumstances of deprivation and torment (Sacks 2002). They were older than me. If I felt the same, the feeling and memory have been lost in the mists of time.

I was sent to Clare Park School in Crondall, near Farnham in Surrey, close to the Hampshire border, and stayed there from age four to eight. It was a three-story building with white walls, surrounded by lawns, flower beds, and fields. A school for girls that catered to the needs of parents overseas, it was possibly coerced by the government into taking in about a dozen boys and a dozen extra girls from the London area at government expense. Perhaps I was chosen because my parents were divorced, and I was without a caregiving mother. School reports (1943–44) were sent to my father. I have a recollection of each parent coming (separately) to see me. My mother reminded me that when she said goodbye, I showed no sadness, only great excitement, dashing off to play with my friends and new stock of toys.

The school was run by a Mrs. Mabel Scutt and four Scutt daughters, names that could have come out of a Dickens novel. I remember the first meeting with two of them, probably Mrs. Scutt and one daughter, Ethel, who looked just as old as her mother in my juvenile eyes. This was in the "Drawing Room" to the left of the entrance, a large room with a tall ceiling, comfortable upholstered chairs, a piano, and a fireplace, yet forbidding in its strangenesss and associated with apprehension in my memory. The room to the right of the entrance, a schoolroom for eight of us boys, has happy associations. I have the odd memory of being transfixed by the color "yellow ochre" from a box of paints that I had just received—Mother's gift, I believe. This room was where I had the rare thrill of receiving a letter.

I was a mischievous boy. Told not to get out of bed by one of the stern Miss Scutts, no sooner had her footsteps receded than I was out of the bed again, only this time the creaking floor that signaled her approach did not warn me early enough. I was caught, and she gave me a stinging whipping with the back of a hairbrush. The paraffin lamp flickered a pretty pattern of light and shadows on the ceiling in the dormitory, creating a dancing light that diverted my attention away from the pain and offered me solace.

I had my friends, but I had my enemies too. The principal one was a nasty piece of work called Peter Rathbone (he may be nice now). He had his devotees, and they took great pleasure in goading me with insults. One afternoon they went too far. I was incensed but cool enough to hold back and play innocent until Peter Rathbone was close by, when I suddenly leaped at him, knocked him to the ground, and pummeled him almost senseless. I can remember the rage mixed with shame, guilt, and a few tears, as breathing deeply and with chest heaving, I got up and walked away. None of his disciples dared touch me, either then or thereafter. I have never done anything remotely like that since. Thomas Huxley describes having the almost identical experience and outcome (Huxley 1900, 5).

I must have been lonely and homesick at times, because I once embarked on an escape—this time absconding without a girl. Perhaps I was six. The great escape must have been when I was old enough to think strategically about where the train station might be and how I might get there but too young to realize it was more than two miles away, I needed money to get on a train, and I had no idea where home was. Also, I might have been wrong about the direction of the station. I walked down a long lane around the back of the school with great determination and reached a gate. I can almost picture it. The gate would not have been difficult to climb, but I must have become demoralized by seeing nothing beyond the gate except a winding path in the dark woods. My confidence evaporated, and I returned crushed by failure and miserable. I told no one and never repeated it.

On the other hand, I have joyous memories of Clare Park, out of doors and on my own, chasing butterflies, for example, and trying to catch a blue butterfly by hand in the dell, my heart beating fast. The dell was a pit in front of the school that I believe was formed by a bomb dropped by a German plane returning to the continent after a night raid early in the war. It was lushly vegetated with grasses and flowers. Burnet and Cinnabar Moth larvae made cocoons for pupation on grass stems, and they glistened in the sun, half silver, half gold. Enrapturing! I was mesmerized by swallows swooping low over a field to feed close to the ground after the rain had stopped, and I have an almost photographic image of the birds, the field, an iron fence at the back, and solitary or small groups of tall, majestic oak trees in full foliage in the background. I stood as still as a post while the swallows swirled around me.

Then there was the exciting occasion when a large boy, perhaps a nephew of the Scutts, came one summer's day and took a group of us out into the oak wood at the back of the school. I saw a Song Thrush or Blackbird nest and wondered in vain how I could possibly climb the branchless trunk to reach it. The highlight of that day was the capture of a black snake, probably an adder. Our big leader cut off its head with a large knife, and the red blood against the black skin has proved to be an enduring image. I was probably shocked, scared, and simultaneously thrilled.

In all these experiences I became more and more imprinted on the natural world. A persistent yet unanswerable question I have had ever since is why I was touched so deeply by nature, whereas every other boy and girl at the school was not. Is there a neurological wiring that makes some of us especially likely to respond to the stimuli of nature as we develop? My father and mother gave me no direction or encouragement, as neither of them had the slightest interest in the world of animals and plants. Cousin Brian played a large part in fostering my intense interest in nature; nevertheless, I believe it resided in me even before that.

Early on I contracted hepatitis. My tonsils and adenoids were taken out, as this was the current practice, and I was confined to bed. Eventually, after a period of drinking without eating, I was allowed to have a nice soft Victoria plum. The experience was not a happy one, because even so slippery an object as a plum was almost impossible to swallow, as it made my throat so sore, but it was better than the accompanying tapioca ("frog spawn"), which I hated and have never eaten since.

On a happier note, I remember wearing baggy brown Polish trousers and dancing the polka with a young girl at the annual summer sports day in 1944. I played the triangle in the band, won the sprint race of eighty yards and the egg-and-spoon race, and had fun in the three-legged race and the sack race.

This far from idyllic but protected existence came to an end some months before the war was over. We evacuees all went home at Christmastime in 1944, when it seemed as if the war would soon be finished (Fig. 2.2). I had been home during summers and have photographs that show I went home, yet strangely I have no recollection of summer visits, and the photographs themselves trigger no deeply dormant memories.

Years later, when on sabbatical leave in Oxford, Rosemary, the children, and I drove down to Surrey and found Clare Park still standing. At a distance it seemed that nothing had changed. We heeded a "Private. Keep out" notice and didn't drive down the long gravel approach to the front door of this Georgian mansion. Then in March 2014, we returned again; this time Rosemary and I were driven by my sister Sarah* and her husband, Alex, and we made it to the front door. It had been converted to a retirement home and no longer stood alone but next to a modern orthopedic hospital in well-designed low buildings. We got out of the car and walked a bit. Had I been on my own I would have rung the bell, announced myself as an "old boy" of the "old school," and asked to take a peek at my painting room on the right. From the outside, the building was just as I had remembered. The dell where I had caught butterflies and watched newts was nowhere to be seen and must have been filled in.

---

* Strictly a half sister, as we have different fathers.

FIGURE 2.2. Looking up, at about twenty-three and five years of age.

Likewise, a large ornamental clump of Pampas Grass was missing. The upstairs bedroom where I had been spanked for getting out of bed could only be imagined, and the same was so for the curved path at the back where I had walked on the failed great escape. But the oak wood was still there, and I could identify the place behind the building where I had lost my temper and attacked the obnoxious tormentor Rathbone.

More than fifty years after I left Clare Park, Rosemary and I spent an evening on Roosevelt Island, next to Manhattan, with the British ambassador, Sir John Weston, and his wife, Lady (Sally) Weston, and our daughter Thalia's husband, Greg, who had guided the Westons in Galápagos. We were the guests of our Princeton friends Richard and Alison Jolly. It turned out that Sally, Richard, and I had been evacuated during the war. Interestingly, our experiences could not have been more different. Richard was sent to a foster family in Ontario, Canada, and bonded so closely they became his parents more than his real ones ever were. Sally, on the other hand, was evacuated to a poor Welsh family with almost a dozen children, where the punishment for disobedience was being shut in a pitch-black cellar all day. In spite of the differences, there was a common denominator. We all commented on how we had hardly talked about our experiences until, strangely, just recently at the age of about sixty. All three of us had many friends but few close ones apart from spouses. These seem to have been common features of evacuated children separated from their parents (Wicks 1988). I interpret them as a suppression of traumas, traumas that influenced our developing personalities but did not impede successful careers.

When I left Clare Park, home was 6 Fairfield Close in Shirley, one of six semidetached houses at the end of a long road. It lay about twenty yards inside the Kent boundary from Surrey and was owned by Auntie Win, who was away in Nelson, British Columbia, with her husband, Vic, and newborn, Lynda. My grandparents shared the house with my father. Surprisingly, I remember nothing of Christmas.

It might have been premature to send me home. At Clare Park I had often heard warplanes at night, seen Heinkels and Messerschmitts, Hurricanes and Spitfires in daytime, and watched chases and dogfights in the sky. I sent my father a letter with a drawing of two Spitfires, and a Messerschmitt hit by antiaircraft artillery fire (Fig. 2.3). Now I saw the planes again, as well as the dreaded buzz bombs or doodlebugs (V-1 rockets). The pulsating whine of these was frightening, but the abrupt stopping of the noise was even more so, because that was when the rocket ceased flying and headed for the ground. V-2 rockets were worse, as they flew at the speed of sound. One night I was woken up and hurried into the air-raid shelter that was half-buried underground in the garden and went to sleep again. That night a V-2 rocket scored a direct hit on the local chocolate factory. I knew there was a factory a couple of miles away but didn't know that "chocolate" was a code name, verbal camouflage for the explosives manufactured there. A large area around the factory was devastated. The explosion blew out the windows of our house, or I should say blew in, because the next day the house was a mess, and I was not allowed anywhere near it. I probably stayed in the comfortable, even cozy shelter. My father was an air-raid warden, having been found physically unfit for military service, and that night was terrible for him. I overheard a fragment of his conversation about covering up mangled bodies with blankets.

This was one of the final attacks, and a couple of months later the war was over. Nonstop descriptions of the celebrations poured out from the

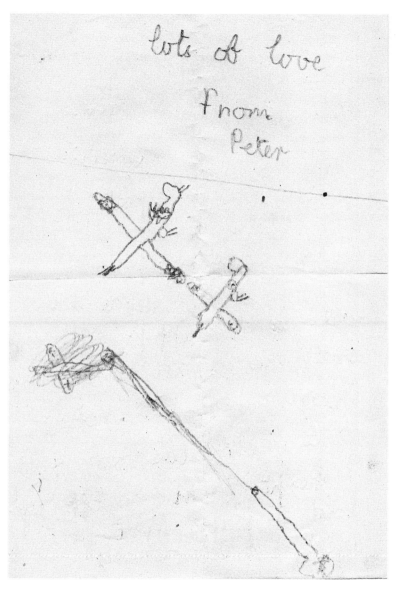

FIGURE 2.3. Artillery shooting down German plane, from a letter to my father, April 30, 1944. "Daddy, wasent there a lot of plaines last night" (letter, April 21, 1944).

radio. Vera Lynn, everyone's favorite, sang, "I am going to get lit up when the lights go on in London," and everyone did, in Piccadilly Circus, which we all went to see one exciting evening along with a few thousand others. The newspapers, the *News Chronicle* in our case, had published awful pictures of the war up to that time and now started with stories of prisoners, concentration camps, and other horrors revealed when Germany surrendered. I do not know how many thousands of children died in the war, but ever since my teenage days I have been aware of my luck in having been missed by the bombs; none of the bombs "had my name on it," as we used to say with grim humor. At some time after the war, all of us surviving schoolchildren received a letter from King George VI. Recognizing our hardships, the letter called upon us to be proud of our country and finished with the admirable wish that when we are grown up, we will "join in the common effort to establish among nations of the world unity and peace."

That first summer was wonderful. In retrospect, it seemed especially warm and sunny, and so were the moods of people, owing to the enormous relief at the ending of five and a half years of dreadful war in which everyone had lost a relative. In our case the victim was Uncle Alan, Auntie Win's first husband; a fireman, who lost his life very early in the war when a burning house collapsed on him as he was trying to save the occupants.

The garden was my patch of flowers and butterflies. In addition to the wonderful colors of butterflies, discovering where Cabbage White butterflies laid their eggs, and picking runner beans and peas for the kitchen, two memories of the garden stand out. The first was climbing out. Right by the fence in the neighbor's garden were some tantalizing stems of rhubarb, some green and some red. Well, the temptation was too much to resist, and though fearful of being caught, I climbed over the fence as fast as I could, snatched a couple of stems, and was back over the fence in no time. My crime went unpunished, and the trophies were delicious. Ever since, I have loved eating raw rhubarb stems, no matter that they make my teeth go on edge. The second memory was somewhat similar, a theft of gooseberries when nobody could see me—except that Auntie

Win did see me, from the kitchen window. She was often a stern disciplinarian, and positively frightening while telling me she knew I picked gooseberries because she had counted them on the bushes. *She couldn't have*, I thought. *It is impossible to count them all.* But I didn't dare take the risk of being wrong. She was merciless in teaching me not to say "wot" but "what" when I was asked a question. I must have said it a few million times and been chastised the same number because the nerve trace in my brain is very strong!

I contracted whooping cough, probably chicken pox, and I don't know what else when I stayed with the Jackman family. At school I caught measles and took it home. On one memorable day I lay in bed in a state of semi-euphoria listening to the rustle of Lombardy Poplar leaves with sunshine streaming through the bedroom window. They were so loud! I was emerging from a horrible illness that had laid me up (or down). I don't remember the fever or anything else about it, only the emergence from it into lovely sunlight, like being reborn. I can still hear and see those leaves on the tall trees, flickering in the breeze.

On one occasion our neighbor Lilian Smith stayed the night as a sitter. A strange woman, she had been recently divorced from her husband, Bill. I woke up when she climbed into my bed, and I shot out like a rocket, probably with a shriek. She disappeared. Next day I told my father, who was furious, and none of us ever spoke to her again. Poor woman; looking back as an adult I think she must have been lonely, and like many primates she craved physical contact.

The few years after the war were a period of food shortage at times even more stressful than during the war. Rationing of items like butter and sugar continued until 1954, when I was seventeen. My father's insurance company received food parcels from an affiliated company in Australia, and they included luxury items like tinned fruit. For several years we lived under the constraint of saving electricity by rationing hot water for our weekly (Saturday) bath to the strictly regulated depth of four inches! The first bananas and first oranges were excitedly heralded by the adults as a return to prewar normality, just as other aspects of that normality were beginning to fade. The horse-drawn carts that delivered milk in bottles, and others that brought coal, were about to be replaced by electric vehicles; and paper tissues, plastic bags, television, and antibiotics were yet to come.

I spent two years at Winton House School in Addiscombe in South London, just a couple of miles from home. I can remember the old, undistinguished, three-story, dark brick schoolhouse, as viewed from Addiscombe Road, and little else except for an intertwined *W, H,* and *S* in purple to form a badge on my pocket, a vague sense of the roughness of the boys, and no recollection of the masters whatsoever. Did I learn anything? Maybe. School reports paint the same picture as the ones my father received in my last two years at Clare Park. Apparently, I showed interest in arithmetic, geography, history, French, Latin, and natural history but was "inclined to be lazy and inattentive—He is bright and intelligent but rather thoughtless. His conduct and general work will both be good when he learns to concentrate and think." This was true; daydreaming was my forte. "He is always cheerful, and is a keen sportsman" (1946).

My memories of that period are centered on home, where my life revolved around the new pleasures of picking gooseberries and currants and catching Common Blue and Small Copper butterflies in the buttercup-filled garden, and rare holidays on the coast (Fig. 2.4). I was given freedom to explore, locally, and took it. I discovered I was fascinated by the variety of different cars in the neighborhood—Morris, Austin, Standard, Hillman, Rover, Sunbeam-Talbot, Wolseley, and many others, all with their unique emblems of identity. Equally, I was fascinated by the numbers on license plates, which I wrote down in a little notebook, and patterns in the numbers. From cars I moved on to do the same with bus numbers and then train numbers. Trainspotting was a sport enjoyed by so many children that books of the respective locomotives were sold, to be used for cataloging and archiving one's captures. A stamp-collecting hobby followed later. At about this time I also learned the pleasures of bouncing a tennis ball or testing my control by aiming and throwing it at an object. These two activities were pointers to the future, innocently expressed from a deep neurological wellspring and not taught by an adult as something that was good to do. Numbers and the diversity of

FIGURE 2.4. **Left**: With my father at Bournemouth, UK, 1947. **Right**: At the same place.

things, be it cars, postage stamps, or butterflies, later led me toward science, and tennis balls steered me toward sport.

During those years, Brian and I were as thick as thieves, except when it came to the Boy Scouts. Brian was a Cub Scout; I was not, as I had no liking for their group activities and enthusiasms. Together we spent part of two summers with Uncle George and Auntie Vi at their large solitary house in Darley Dale, Derbyshire. Brian had been once before and had befriended Norman Stone, a local boy of the same age, and Norman introduced us to the local dialect ("Goin' up clough?"), to local lore, and to tickling trout in a small tributary of the river Derwent. I tried it and failed to tickle even one, at a site just across the river from where otters had a slide. I never saw the otters themselves. We went up to the bracken and heather moors, looked for exotic caterpillars of Emperor Moths, and saw a Ring Ouzel. We were explorers, cowboys and Indians, trappers. But then we turned into geologists and scoured a mine tip for fragments of the colorful crystalline blue john, a rare fluorite sometimes known as Derbyshire spar. It was on this tip one cold December day that we tipped our arrows with the spent bullets of rabbit hunters, pulled on our mighty ash bows like Robin Hood (Fig. 2.5), and nearly hit our targets—greenfinches! In all this, my early imprinting on nature, first at Nonsuch Park and then at Clare Park, received a huge reinforcement.

When confined to the house one wet day, Brian and I made maps of the significant landscape features from the village up to the house: the hedges, the fields, and the bends in the narrow road. About ten years later I returned to Darley Dale on a day's break from national service in Cheshire, and I was amazed at how short the actual distances were compared with those in the maps and my memory. Another relic from that period I have treasured is a Clouded Yellow butterfly* that I caught in the grounds of Nonsuch Park after an enormously long, torturous, and anxious chase. I can recall the sandy bank where it finally settled and I caught it (by hand), and the feeling of triumph with the poor butterfly sealed in my closed hand. This was the glorious summer of 1947, when butterflies were everywhere, many having flown over from the continent. It never rained!

Several years later, our interests took us on diverging paths from an initial, shared fascination with nature. Brian had a successful career in journalism and writing, specializing in travel and nature walks, and was a pioneer of ecotourism in Africa (Jackman 2021). Adjectives attracted Brian; numbers attracted me. Our divergence was already evident on one of our winter bird-watching trips to the Barn Elms reservoirs in North London to look at migrant ducks. As a flock descended from the sky, Brian imagined painting them against rose-tinted clouds in the style of Peter Scott and suddenly became aware that his companion was saying "twenty-four, twenty-five, twenty six . . ."!

Auntie Win came home a year after the war ended with her Canadian husband, Uncle Vic Thompson, a soldier, and their two-year-old daughter, Lynda. He gave me a jackknife that I kept for many years. I suspect he was there for divorce proceedings, because he left after a short stay, never to return. A vivid memory of that time, because it was often

---

* It is now in the American Museum of Natural History.

FIGURE 2.5. With cousins Jennifer and Brian, as "archers of the glen," at Warren Carr, Darley Dale, Derbyshire, UK, 1946.

repeated, was accompanying Auntie Win and Lynda to Beckenham or Penge to collect the government-issued weekly child allowance. There were ration books, green ones and blue ones, with little tickets that were exchanged for sugar, butter, and orange juice. I remember the orange juice because I wanted it, but it was destined for Lynda, and I never received so much as a sip, and rightly so.

Eventually Auntie Win decided to sell the house, and that meant we had to move. The reason was a third marriage, to Bill Lorimer, later to become Uncle Lorry. I was all set to go to Beckenham Grammar School, and Nan (my paternal grandmother), bless her, gave me a bicycle as a reward for getting a scholarship, and later she gave me my first watch as a birthday present. But we moved out of the area, and I never went to that school. My father found us a flat above Mr. Miles's Car Repair shop in Norbury on the London–Brighton Road. Our address was 1150a, and that is where Dad (at that time Daddy), Nan (at that time Nanny), and I relocated. I exchanged a garden of buttercups, fruit, and vegetables for a backyard of old tires and filthy oil. But there was one compensation, a very small unused space, three roofless walls surrounding a ten-by-twenty-foot floor of flat concrete with occasional weeds in the cracks, where I could practice cricket for hours on end by throwing a tennis ball against a wall and hitting it on the rebound with the grace and power of my England cricket heroes Len Hutton and Denis Compton.

Across the road was an unoccupied bomb site strewn with brick rubble and overgrown with *Buddleia*, well named as butterfly bush, whose flowers were a powerful attractant to butterflies of all types. Nobody troubled me; I treated it as mine, my own garden. But on one occasion, much later, when I was seventeen, an irate dentist emerged from an adjacent house, swore at me for damaging his plants, came over, and cuffed me hard on the back of my head. It hurt, but much worse was the indignation of being punished for something I didn't do. The unfairness of it! Sobbing, I told my father, and he went flying across the road to confront him. I had never seen him so angry. I thought he would "fetch him a fourpenny one," to use a cockney expression for hitting him. Fortunately, he didn't.

3

The Gift of Whitgift

THE IMMEDIATE problem was finding a school for me, even before the move to Norbury. My father took me to look at two schools. The first was John Ruskin School, and the second was Whitgift Middle (later renamed Trinity School), in the middle of Croydon and founded in 1596 by John Whitgift, archbishop of Canterbury. I was scared by the boys at both places; they were rough and tough, and the buildings and masters were also intimidating. I was miserable. With great reluctance, my father then took me for an interview at Whitgift School in South Croydon, a fee-paying public day school that had been split off from the middle school and relocated in 1931 to Haling Park (Percy 1991). Haling Park, like Nonsuch Park, was once an estate of King Henry VIII, and later it was the home of Lord Howard of Effingham, a high-ranking diplomat and admiral in charge of the military response to the Spanish Armada in 1588. An imposing brick building, with a replica of Lord Howard's galleon, the HMS *Ark Royal*, above the building, the school stood on a hill overlooking a playing field and the main London–Brighton Road and had about eight hundred boys from age ten years and up (Cox 2013). I immediately felt the school was the place for me—the atmosphere was calm and respectful, the playing fields were large, and it felt like being in the country.

I managed to pass the interview, and although I was not good enough to qualify for a scholarship, my father was offered a grant from the Croydon County Council Education Department to help cover the costs. My father acquiesced, and I shall be forever grateful (Fig. 3.1). He was

FIGURE 3.1. First year at Whitgift School, South Croydon, 1947.

reluctant because of the cost, and many times later reminded me of what he had to give up for me to receive a superb education. I am not arrogant and could never have been, with a father like him. There was a social dimension to his reluctance too, as the school looked as if it was for the rich, and he felt ill at ease in that environment. As I later discovered, he was particularly embarrassed by his London accent when meeting any-one associated with the school. Mine was probably the same.

To put my education into context, we were socially and economically at the low end of middle class. My paternal grandfather was a Royal Marine in World War I and a postman; his father was also a postman, and his grandfather was an agricultural laborer. Lateral branches of the family in the nineteenth century were rural folk employed as builders of houses, thatchers of roofs, and coopers of barrels. My paternal grand-mother was raised on a farm in Potton in Bedfordshire, and at the age of thirteen she was in "domestic service" in London, scrubbing people's steps and working in a kitchen. My grandparents had almost no education and not much money, and yet they knew the value of education and paid for their four children to have piano lessons, which I find astonishing. My maternal grandfather was a chemist for a company that tanned leather; he lived with my grandmother near Beckenham in Kent. They worshipped my father and despised their daughter for abandoning him and me. My favorite memories of weekend trips with my father to see them were visits to the wartime allotments where Grandfather grew potatoes, runner beans, peas, and other vegetables on a small plot of land complete with a small potting shed for tools.

My father had left school early, trained for nothing in particular, and had menial jobs until he was thrown out of work in the Great Depression in 1929–30. When, eventually, he managed to get a job, it almost broke his back, literally. The work entailed hauling frozen carcasses of beef on his shoulders and back from a frozen locker and tipping them into the rear of a truck. He had a slender frame, and when his back did indeed give out, he was once again out of a job. A minor clerical position at the General Accident, Fire, and Life Assurance Corporation (GAFLAC) saved him from penury. It was well below his abilities, but it was secure, and that was very important in those uncertain times, so he stuck to it for the rest of his life. He read novels and newspapers but little else. We had few books in the house.

My father played a key role in my development, not only paying for the education despite a low income but encouraging me in all ways. My two passions were natural history, mainly butterflies and birds, and sport— cricket and hockey. Academically, I was right in the middle. Each cohort was split into four groups of twenty-eight. I was just under eleven and the youngest in my class when I started in the lowest of the four. In the next year I was promoted to the third-ranked group and was one of the top three boys, and that is how my academic position stayed until I reached the unstructured sixth form.

Biology was my main passion, ahead of mathematics, languages, and English grammar. My father had a similar fascination for our own language and shared it, so not surprisingly it was one of my best subjects. Trigonometry and algebra were fun because they were problem-solving exercises. I loved French and then later German, and the imitation involved in learning how to speak them. English literature, chemistry, physics, history, and geography lagged behind the rest, in part, but only in part, owing to less-than-inspiring teachers.

I had to repeat a year for the strange reason that to take the General Certificate of Education (GCE) exams at ordinary (O) level, we had to be sixteen on September 1, and I fell short by several weeks. I felt I was held back and simply marked time during that repeated year. Eventually, I managed to pass all subjects except for English literature. There is a reason for this. We had a school cricket match at the time the exam was to be held, and I was allowed to take the exam immediately beforehand, when my mind was not wholly focused on the task at hand.

I suffered a big disappointment on entering the lower sixth form when I ran into difficulty with calculus. Mathematics had been a favorite subject, and in the fifth form I had received a mathematics prize, but here I was suddenly struggling. The problem was that the teacher, a man called Harry Stothard and nicknamed Gloom (!), went too fast in his lessons, and once behind I had great difficulty catching up. My father could not help with my lack of understanding. In contrast to that experience, I had no problem with the mechanical world of physics. Biology, though, was my major interest, and it was taught excellently by Cecil Prime (botany) and Bob Jones (zoology). Both put fuel on the fire and fanned the flames of my expanding interest in biology and the natural world.

They did more than teach; they arranged for a group of us sixth formers to attend winter evening lectures at the Royal Institution in London. I must have been enthralled by at least a half dozen, some of which stand out in my memory. H. Munro Fox gave thrilling lecture on animal colors and pigments that opened my eyes to a new world, as good science should do. I became fascinated by the chemical structures of porphyrins, pterins, and carotenoids, their differences, and how they worked together to produce many of the colors we see in the animal world. Physical demonstrations on the bench in front of the speaker were a big feature of these lectures.

Later I watched an interview on television with the director of the Royal Institution, Lawrence Bragg—a Nobel Prize laureate in physics for his discovery of X-ray diffraction—who was asked if, at his advanced age, he was nervous before giving a lecture. His reply was memorable: "Yes. And the reason is I always want to do the best I can, because the people who come to hear me on a late, cold, winter evening could be doing lots of other things, so I have a responsibility to make it worth their while not to do those other things but to spend an hour listening to me!" His philosophy stood me in good stead as a reminder of that responsibility in later years when I became a lecturer.

Another initiative of our teachers was to invite Jack Lester, the curator of reptiles at the London Zoo, to come to Whitgift and talk about reptiles. I have forgotten the lecture but vividly remember a long boa constrictor he brought. He wanted us to learn how they feel, not hot and slimy as you might think but cold and dry, and to demonstrate the point, he told us to hold out our hands at full length, and the six-foot-long snake made an undulating journey from one boy's hands to the next for the length of the row. I can almost feel the serpentine coldness now.

For the advanced-level biology exam, I did a botanical study on nearby Mitcham Common in the roughly vegetated areas close to a golf course, mapping different grasses and doing elementary soil chemistry to try to account for spatial differences. Nothing adventurous—it was a descriptive, modest exercise in scientific discovery, but I was absorbed in the details. In the end, given an average grade, it helped me to pass the biology part of my GCE, but more importantly, in the long run, it was an invaluable experience in designing and carrying out my own research. Dr. Prime was an excellent, gentle, probing critic who repeatedly made me think again. His biggest gift was in making me articulate scientific questions simply and clearly.

My intense interest in biology seemed to lead nowhere because the career possibilities of medicine, veterinary medicine, dentistry, forestry, fisheries, and agriculture did not appeal to me. Medicine, in particular, was clearly not for me. That possibility was foreclosed when, several years earlier, I nearly fainted at the sight of blood being spilled in a film about abbatoirs. Mr. Ewen, the careers master, was exasperated with me,

I am sure, having failed to suggest anything that interested me other than combining my interests in biology and sport and becoming a schoolteacher. And sport was a major passion, so that did not seem to be such a bad option.

Sports, not scholarship, gained me prominence at school, and I was chosen to be the first captain of a new "house" named Andrew's after a past headmaster. Whitgift was a rugby-playing school, but rugby was too physical for me. When Bob Schad, an international hockey player, joined our teaching staff, he introduced hockey, and I blossomed. I was chosen to be the captain for the last two years. The team was good but not outstanding, and the same could be said for me, the center half. I was good enough, however, to be selected to represent Surrey Juniors in a home-counties tournament, which we won in a memorable hard-fought final game against Sussex Juniors. That gave me the first medal of my life.

One sport gripped me more than hockey, and that was cricket. Here again, I was prominent, finishing up in the final year as captain of the school team (Fig. 3.2). I was an opening batsman and occasional spin bowler. One match for the school team was watched by a reporter from the *Daily Telegraph* newspaper, and in the next day's issue he reported that "Grant has a nice pair of wrists," referring to how I handled the bat—a huge (national) compliment. My golden memory is scoring one hundred runs against a team made up of the masters of the school. I had made ninety-three by the tea break, and on resuming they made a very determined effort to get me out, with Eddie Watts, formerly a Surrey County bowler, doing his hostile best, as I crept to one hundred in singles. In athletics I had stamina but not strength, could not run very fast but could run long distances, and was third best at running the mile on the school team.

Returning to the question of a career, I was given advice to apply to universities because my sporting achievements combined with an average academic performance would give me a good chance of admittance. This was out of my father's realm: no member of our extended family had been to a university, and we scarcely knew anything about them.

FIGURE 3.2. As captain (front row, center) of the Whitgift School cricket team, 1954–55. (Two members are absent.) Jeremy Evans, on my left, is mentioned on p. 29; and John Webb, on my right, on p. 34.

Following advice from the school, I applied to the three London colleges—King's, University, and Imperial—and to my great surprise was accepted at all three. However, I did not go to any of them, thanks to an intervention from Bert Parsons, a geography teacher. We had made initial contact through sport, and it was he who told me, nicely, that I was not powerful enough to be a speed runner because I did not lift my legs high enough. True, I loped more than galloped.

He suggested I should apply to Cambridge University. This presented a problem—I would need to pass Latin at GCE ordinary level, but although I liked Latin and performed well, I had given it up several years before. Behind the scenes, Mr. Parsons had talked to Charles Whyte, our classics master, and this generous and gentle man offered to coach me at his house on Saturday mornings. He and his wife were childless, and so extraordinarily kind and generous they deserved to have children, as they would have been very caring parents. Anyway, I passed when exam time came around at the end of the year, though at the lowest grade possible. Mr. Whyte gave me a silver pencil with engraved initials as a present for passing. I hope I gave him enough lively and amusing conversation in return. I had to be careful not to make him laugh because his laugh was like the whicker of a horse!

At GCE advanced level, I passed biology comfortably, chemistry with little room for error, and physics by a generous decision on a borderline result. Three A levels and Latin at O level were requirements, and I just squeaked by. On such paper-thin margins of error (or safety) are careers built. We can all tell stories of "if/if not" moments in life. For me, getting in to Whitgift was one; admission to Cambridge was another.

Our headmaster, Geoffrey Marlar, contacted his college, Selwyn, and recommended me. The college has a strong tradition in religion, having been founded by Archbishop Selwyn in 1882, and I was asked whether I was baptized and confirmed. No, I wasn't. At the age of seventeen, I was in a religious phase, wanting to believe in the fascinating miracles of the Bible, such as Jesus walking on water. I had won a prize much earlier with an essay on the miracles, so I felt it was natural to receive some more coaching from within the church, and I was confirmed as a member of the Church of England. By the time I reached Cambridge a couple of years later, my religious phase was over; I had become critical of myths and the supernatural, and I was on the way to becoming an evidence-scrutinizing, empirical scientist. My future mentor at Yale University, G. Evelyn Hutchinson, described himself at the same age with words that resonated with me. He was, he wrote, an "English countryside empiricist" (Hutchinson 1979, 56).

These reflections are dominated by the successes, failures, and activities of the later years. A different narrative would begin with my first day at Whitgift School when, in my excitement, I lost my running shoes for the gym. "How?" asked my father sternly that evening. I explained I had been told by another boy to put them on top of a locker before class, and they were gone when I returned. I then received a lesson in looking after myself. "If someone told you to put your head in a fire, would you do that?" Not surprisingly, I learned the lesson instantly and forever. Not only that, I have a very vivid memory of the green lockers, their exact location on the second floor of the junior school, and where I put the gym shoes—pedagogical gym shoes one might say. Other boys have told of similar traumatic experiences on their first day at Whitgift (Cox 2013).

In that first year I had to learn how to survive the hostile social world of ten- and eleven-year-old boys in the playground. Physically a shrimp, but an agile one, I learned to use wit and nimbleness to avoid the few bullies and tyrants among us. In the classroom I could not control my rebellious, mischievous behavior. Twice I was beaten. The first, by the teacher Freddie Percy (nicknamed "Chin") and fully deserved, was with a broken-off piece of blackboard frame, sporting a memorable nail marking the base of the "handle" end. The second was completely unfair. Bill Edge, teaching us history, could not control his class, and every time he turned his back to write on the board, we started talking. He would then whip around to try to catch one of us, yet rarely succeeded because we instantly fell silent. On one occasion he had worked himself up to quite a pitch, and turning from the board he shouted "All right, I have had enough of this. I am going to cane the lot of you, starting with you, Grant." (I was sitting at the first desk on the left of the front row.) I was instantly angry and indignant because I had not been the culprit on this occasion. I walked out, bent over, and *whack* came the cane, followed by the second blow. But I had already started to move, and the third strike barely touched me as I sprinted out into the corridor followed by the irate teacher. Worst of all, Bill Edge calmed down and caned no one else. I crept back. This was the time to heed the school motto, *Vincit qui patitur* (he who suffers, conquers), since I had failed to remember my Clare Park school motto, *Vincit qui se vincit* (he who controls himself conquers).

When I left school, I was a prefect and second to the head boy, Jeremy Evans. I was sad to leave. I had had such an enjoyable time, and although there was an exciting trip to Iceland to look forward to, I also had to face two years of national service after that.

A humanizing feature of Whitgift was our escape at the end of each day, back into society and home. There was a life in school and a life out of school. On weekends my father introduced me to interesting parts of London, to landmark buildings like Westminster Abbey and Saint Paul's Cathedral, to the British Museum and the Natural History Museum in South Kensington, to Hyde Park, the London Zoo in Regent's Park, Soho, and streets with interesting names and histories, like Pettycoat Lane, Tongue Alley, and Glasshouse Street. One of my favorites was the National Gallery, and I returned several times during my teens. It was easy. I got on the number 109 bus at home, got off at Trafalgar Square, and walked across to the gallery. The visual arts are a natural attraction to a budding naturalist, since observing, identifying, and interpreting are shared skills in the domains of culture and biology. Changes in modes and styles of painting, and the replacement of one by another, are like evolution in the biological world—extinction of organisms and replacement of some by others.

My father educated me in painting traditions and history. I started, naturally, with landscapes by John Constable and the Dutch painters, then became fascinated by the Impressionists and how they achieved their visual effects. In different ways and for different reasons, I loved Rembrandt, Turner, El Greco, Goya. Part of this admiration came from what I learned from our art master at school, Frankie Potter, who had studied at the Slade School of Fine Art and looked to us in the classroom as if he were trying to be Pierre Bonnard, the nineteenth- and twentieth-century French painter, with whom Mr. Potter shared a particular liking of florid reds. What seemed like entertainment more than instruction—we could not resist imitating his stammering and explosive, witty outbursts at some misbehaving student—was actually an educational foundation for my appreciation of the visual arts. I regret that I had no talent in drawing and painting.

At about age fourteen I became more interested in birds and joined the bird society. Our senior members were part of an information network that used the telephone to alert everyone to unusual birds in the vicinity. On one occasion a Great Gray Shrike from Scandinavia was reported to be on Mitcham Common, so off I went on my bicycle to see it. On another occasion there was news of a quail, so again I got on my bicycle and rode many miles to the place, Oxted, I believe. Well, I was lucky, because it was not only calling from a wheat field but got up and flew a short distance before crashing back to earth again. Sixty years later I heard, but did not see, my second one at Guarda in Switzerland, and remarkably, I immediately recognized what it was.

The leader of our bird-watching group was Derek Pomeroy, later professor at Makerere University in Uganda. He had the personality of a Boy Scout leader, organized us efficiently, and took us on day trips to the coast to see shore, marsh, and water birds in Sussex and Kent. Emboldened, we spent a week at Seahouses on the Northumberland coast, where Eric Ennion, a welcoming, most hospitable doctor, ran a hostel for bird-watchers and other naturalists. He was also a superb bird artist in watercolors. The ornithologically rich Farne Islands could be seen at a distance, but the sea was too rough to visit them. However, we did have a long and enjoyable hike up in the wildness of the Cheviots, and there I saw Twites for the first time. They are high-elevation replacements of linnets, their more familiar lowland cousins. Even more strongly I remember the first sight of a Purple Sandpiper on the rocky shore. The fun of bird-watching came from identification and admiration, nothing intellectual.

Looking back on these excitements, I experience a chill when I think of the time a group of us walked far out onto the hard mudflats at low tide to look at a variety of wading birds, only to find ourselves surrounded by gently rising water. The mud became soft. I like to think it was I who hit upon the idea of walking back along the beds of little rivulets that had accumulated small broken cockle shells and thus were harder than the surrounding mud. One of us, Tony Holcombe, nearly fainted in fright, but after a long and anxious walk, we finally made it to dry land.

Expeditions like this were adventurous fun (usually) and greatly rewarding in broadening our horizons. Our two biology teachers took us on one-week courses in April, one to Plymouth in 1954 and another to Clapham, near Malham, in Yorkshire, the following year. We hitchhiked in pairs to both places. At Plymouth, the wonderful world of marine biology was revealed to us on board a research vessel, when a net full of dredged benthic creatures was lifted up from the depths, and a mass of new animals spewed forth and slithered across the deck: polychaete worms, starfish, sea urchins, and so on. Back in the lab we drew them and attempted to identify them with technical books. It was research in embryonic form, and encountering new creatures was exciting— another foundational brick in the building of a career.

At Clapham I extended my growing interest in grasses and learned some basic geology from Bob Jones. The highlight was a visit to Colt Park, a small nature reserve described as one of only two completely untouched woodlands in Britain. When we arrived, it was immediately obvious why it had not been exploited: it is essentially a mass of large moss- and anemone-covered limestone boulders, defying entry to all plows. Contorted ash trees, perhaps oak and sycamore as well, seem to wriggle out between massive boulders, twisting and turning, and then stop, trimmed by winter winds. Wistman's Wood on Dartmoor, the second untouched woodland, has a similar appearance of a tormented landscape. We paid a visit on an excursion from Plymouth. Here the twisted trunks of dwarf oaks manage to escape from between large granite boulders but don't get much taller than humans. A thick covering of mosses and lichens makes the boulders look beguilingly soft and green. Wistman's Wood is a remnant of the ancient forests that extended across Dartmoor before they were cleared by Mesolithic hunters and gatherers. The experience of being in primeval forest made a big impression on me, as it set my mind to thinking about the "natural state" of England, when wolves and bears roamed throughout the country and what is meant now by "natural habitat."

Outside school, one of the large events I recall from that time was the Festival of Britain, held on the south bank of the River Thames in London, a joyous celebration that my father and I went to more than once. This was in 1951. The next year was unforgettable for a completely different reason: the smothering of London by an extraordinarily intense fog mixed with smoke, the thickest smog, as it was called, that I ever experienced. I could not see across the road, could not cross it without hesitant steps, and only knew a tram was passing by its muffled clanging sound. Visibility must have been no more than a couple of yards, and a hand at arm's length could scarcely be seen. This was the Great Smog of 1952, during which thousands of people died. It led to the Clean Air Act four years later and the end of coal burning in London (if only the world could dispense with coal as quickly). No one who experienced the "pea-souper" fogs of the early 1950s can forget their sulfurous smell and impenetrable greenish appearance.

In 1949 I was able to travel to Saint Ives in Cornwall with the Jackmans on their annual holiday. Uncle Stan worked for the Southern Railway, and a perk was a free trip by rail for him and his family once a year. The southwestern county of Cornwall is well known for its china clay (kaolin) deposits and tin mining. For us its greatest value was in its climate— more Mediterranean than anywhere else in Britain—and its natural history. The light was lovely, the moors beckoning, the wind soft and warm, and Saint Ives itself was a quiet, charming haven for painters and potters. I loved the strange Cornish accents. A seller of Cornish pasties came down the street early in the morning selling small meat and potato (tiddie) pasties, crying out, "tiddie oggi, tiddie oggi, tiddie oggi oggi oggi!" Exotic. Perhaps a rural Cornishman would have the same reaction in London to a newspaper seller calling, *"Star, News,* or *Standard,"* with the voice rising at the end.

In 1951 Brian and I returned by ourselves and camped just outside Saint Ives, revisiting some old haunts and adding new ones. We were naturalists looking for new things: birds to see, butterflies to catch, and their caterpillars to find. A view of a gorgeous, empty sandy beach in the Hayle estuary stands out in the visual part of my memory, and the wheezing calls of greenfinches on their display flights from the pine trees above our tent resides in the acoustic part.

My father took Brian and me to Paris in 1950, to the Latin Quarter of the left bank, where we stayed in a small pension just off the Boulevard Saint-Michel. It was my first trip abroad, and very exciting, and gave me the opportunity to exercise my schoolboy French. We did all the tourist things, visiting the Louvre, Sacré-Coeur in Montmartre, Champs-Elysées, Eiffel Tower, Notre-Dame, Jardin des Tuileries, and so forth. My confidence grew as I felt emboldened by being able to handle the strangeness of a foreign country where the language was different and a challenge. Both Brian and I left with an appetite for more foreign travel. Four years later I made a second trip to France, this time with a friend, John Webb, and two of our schoolmasters, Bert Parsons and Bob Schad, in Bert's car, an old Alvis. Our destination was the beaches of Menton on the French Riviera, close to Nice and the Italian border, to enjoy relaxing, good cooking, exposure to French language and culture, and companionship.

The Whitgift expedition to Iceland of 1955 was led by Roy Kennedy, a short, stocky, energetic English master, cricketer, and ultra-keen bird-watcher. I was the botanist. To prepare for this trip, I visited Dr. A. Melderis at the British Museum, and he gave me good advice as well as a request to search for a rare little fern, *Ophioglossum vulgatum*, which grows close to solfataras (sulfurous hot-spring vents); and two species of *Agropyron* grasses. Three other boys covered geology and birds, and one other was a general factotum/administrator. We sailed from Leith to the Shetlands and thence to Reykjavík in late July, and I was seasick. In Reykjavík we stayed at the Salvation Army hostel and met the enormously tall and well-known ornithologist Dr. Finnur Gudmundsson for advice on where to look for the rarer birds. Then we rented an ex–US Army Dodge vehicle that had seen much better days and drove up the west coast and across to Akureyri, Mývatn, and Húsavík.

Iceland is a volcanic, green island with large glaciers. We drove along wide valleys with meadows of lush long grass, past fast pebbly rivers and sheep-grazed treeless mountains. Small farm holdings looked minute against this imposing backdrop and were few and far between. When we were not camping in tents, we slept in hay barns, thanks to the friendliness of the farmers. We were all enchanted by Mývatn (Lake of Midges) and its rich avifauna where, unexpectedly, the notorious and ferocious biting midges in their hordes were plentiful but quite peaceable at 79°F. Midges are not tasty, and to avoid eating them we devised the technique of walking a few paces to dissipate them before opening our mouths for each delivery of food. Roy displayed a side that was never on view at school, acting like an excited young schoolboy whenever we saw a new bird, such as a Gyrfalcon or Harlequin Duck. I switched into my role of expedition botanist and studied the flora of two contrasting habitats, *mó* (heath) and *mýri* (marsh). Unexpectedly, I found a hybrid rush (*Juncus balticus* × *filiformis*) in the presence of only one of the parental species (*J. filiformis*) and much more abundant than it, an example of hybrid vigor. Where was the other parental species? Many years later, hybridization was to become a fascinating focus of my research on Galápagos birds (chapter 14).

One day I took off alone across a moorland landscape to Hveravellir in search of the solfatara where the fern had been found. A stream on my 1931 map turned out to be an uncrossable river, but luckily I found a farmer, who rowed me across and then shyly refused my offer of money. On climbing the bank, I entered a cold, clammy mist that reduced visibility to fifty yards. I walked over moorland, waded across icy-cold streams barefoot, continued walking accompanied by whimbrels and ptarmigans, handed on, it seemed to me, by one mournful golden plover to another every hundred yards. The ghostly shadowy shapes of a pair of Short-eared Owls appeared against a white cloud, then disappeared, when I finally reached a track with one mile left to my destination. But I had run out of time. The initial delay at the river eventually proved decisive, and I returned empty-handed.

One object of the expedition was to reach a lake called Askja, made famous by the fate of a German geologist who disappeared there in the late nineteenth century. One minute he was in the water, according to the surviving companion, who had his back turned to him, and the next minute he was gone, and the level of the water had suddenly dropped. That was the survivor's story, at least, and it had a peculiar appeal to Roy. However, we never reached the lake, as our vehicle became stuck in glutinous black mud when traveling across country and had to be towed out by tractor by a farmer from Svartárkot. The farmer, Hjörtur Tryggvason, and his family were very kind and gave us tea and cake. I was amazed to see row upon row of English classics (in translation) lining the wall: Dickens, Shakespeare, and others. There we turned around and went back to Reykjavík. Altogether it was a wonderful adventure, despite three breakdowns of the Dodge and my failed quest to find the fern or the grasses. I was determined to return.

4

Marking Time
in National Service

ALMOST EVERY boy of my age did military national service. For me it was mostly a waste of time. I grew up a little, I am sure, but in terms of intellectual life and a career it just held me up for a couple of years, a period of marking time for the second time of my life, and going nowhere. Trying to look on the bright side of things, perhaps I benefited by being at university two years later and being just that little bit older. The best part of the two years was my first visit to Spain.

I was conscripted into the army to serve in the Queen's Royal Regiment, an infantry regiment with headquarters in the pleasant town of Guildford in Surrey. Military life was a shock for all of us, from the streetwise petty crooks of East London to the public schoolboys of wealthy privileged parents. In bed on the first night, I listened to the boasting of the "jobs" the thieves had "pulled." In daytime we had entered a world of being shouted at, marched up and down the parade ground, and fed a diet of unappetizing food.

After a few days, those of us who wanted to become officers were examined in an interview. Our commanding officer was a friendly and reassuring man, but his companion adjutant, the main examiner, was not. He was a harsh brute who successfully weeded out the arrogant public schoolboys and distinguished the weak from the strong. I had a streaming cold, and I failed, or rather was given what was called a deferred watch, which meant I would have a second chance later. After the first two weeks of relentless and grinding basic training, off I went with many others to Canterbury for further training. I enjoyed Canterbury when we were let out on weekends, and spent some time in the cathedral, famous for its long history, extending back to the sixth century, when Christianity was introduced to England by Saint Augustine, and for the murder of its archbishop, Thomas Becket, in the year 1170. The cathedral is early Gothic in style and built on a grand scale. Sixteen magnificent pillars flank the nave and reach up to the vaults at a great height. For me this quiet and tranquil space was a place for calmness, where I could relax, admire its architecture, and reflect on its history and continuity with my school, Whitgift.

Another source of escape from the military routine was hockey. Eventually I was able to take the officer's exam again. This involved leaderless teamwork, in which collectively we had to erect a large pole vertically so that one person could climb on the shoulders of the others and thence shimmy to the top. I knew what was required to impress, took the lead, and was just about to reach the top when the group lost control and both pole and I crashed to the ground. A nosebleed ended that exercise, and of course I "soldiered on." The upper lip remained stiff, even though the crooked nose betrayed me. Anyway, I passed the exam by showing "initiative" and "leadership quality." The nose has continued to be prone to bleed throughout my life, despite the valiant efforts of an excellent physician in Ann Arbor, Michigan, to straighten it some thirty years later.

The next phase of my national service was spent in officer training at Eaton Hall in Cheshire. The hall was owned by the Grosvenors, the family of the Duke of Westminster, and had been in their keeping since the fifteenth century. The building itself was much younger, built in the nineteenth century to replace a predecessor, and was imposing only in its size and position, surrounded by gardens. It was "loaned" to the army for a period of time, and less than ten years after I was there, it was demolished, and another was built in its place. This much more pleasant existence lasted sixteen weeks, and I "passed out" as number seven in an intake of about twenty-eight, quite an achievement for a non-militaristic person.

On one free long weekend in July, I hitchhiked to Church Stretton, then on to Ludlow, a small, attractive Elizabethan market town once protected by a much older castle. I set off from there to spend an exhilarating couple of days climbing hills, collecting ancient rocks from the Cambrian and Ordovician periods, and exploring the high moorland of Long Mynd, without meeting anybody. I stayed overnight in a nearly empty youth hostel. The day alone on the moorland in Iceland had already given me a pleasant taste of solitary exploration and being a naturalist, I loved this one too. On the return to Chester, I stopped in Shrewsbury. Groups of swifts flew screaming down by the River Severn as I looked up to see Shrewsbury School on the hill. Charles Darwin attended the school and was not impressed: " The school as a means of education to me was simply a blank" (Barlow 1958, 5).

At week twelve we had to declare an interest in joining a regiment, and I put my name down to be seconded to the King's African Rifles in Nigeria. I was very excited at the prospect of going to the tropics and was already looking forward to the weekends off to explore nature, when the Suez Crisis erupted with Israeli–Egyptian hostilities. Britain supported Israel, and I was sent to the movement-control unit in Botley, on a hill above Southampton in Hampshire, with the galling task of sending others abroad instead of traveling myself.

Eventually the Suez conflict ended, and my second year of national service was uneventful. On one weekend in May 1957, I made my first visit to the Isle of Wight, timed to coincide with the flying season of the Glanville Fritillary, a butterfly that has been studied intensively elsewhere in Europe (Hanski 2016) but, strangely, is found nowhere else in England. It was excitingly new to me. My companion was a Major Cowan, a soft-spoken keen butterfly collector, representative of a curiously strong tradition that combines military service with natural history collecting. For example, he had amassed a very large collection of butterflies while serving in Malaysia.

Some of the officers who were my colleagues were very bright and could have had successful careers outside the military. One was Bill Penaluna, an Irishman and an adjutant. He was a great help when I had an argument with a new, very unpleasant commanding officer (Big Jim) about my leave. I had missed the annual two weeks leave in 1956, so I petitioned for a carryover to give me four weeks in 1957. It was granted but only because of Bill. The long time enabled me to escape to Spain for a month.

Cousin Brian and I traveled to Barcelona by train. The country and language were new to both of us, but we immediately felt the warmth and friendliness of a people and culture that has since influenced both of us in our separate lives. We visited the Sagrada Familia, Las Ramblas, and Montserrat, where we established a lasting friendship with Aromas del Montserrat, the liqueur made in the monastery. Brian had read a lot about the romance of bullfights, so we had to go to one. He knew all the technicalities and gave me an education as we watched the spectacle of Carlos Arruza, a famous matador astride a horse, tormenting a bull and then finally delivering a coup de grâce to the exhausted animal. Yes, it was an exciting drama, but more thrilling to Brian than to me, and he was disappointed that I did not want to accompany him to see another the next day.

From Barcelona we traveled by bus to Rosas on the coast, via Figueras.* While we were waiting at Figueras, a group of three modern troubadours got on the bus and entertained us with their singing accompanied by guitars. They were loud and exciting, utterly different in spontaneity from anything in England, and our blood was racing as we headed to the Costa Brava. The week there was splendid. In the bright light I began to appreciate the stark contrasts and crystal clarity of the Salvador Dalí paintings I had seen under artificial lighting in the National Gallery in London.

Brian left Spain after our visit to Rosas, and I headed south, both hitchhiking and taking trains. In Madrid I stayed in a pension run by a friendly barber and his wife. I was a tourist again and enjoyed walking around, sightseeing, and visiting the Prado Museum. I would like to forget the postcard I sent home excitingly declaring the paintings were

---

* On my visit in 1957, only Spanish was used. In fact, Catalan had been outlawed by dictator General Francisco Franco. The law was relaxed or overturned only after his death in 1975.

all originals. I have no idea where I had got the idea they might not be. I seem to have been more naive than I wish to remember. Twice I was duped in Madrid. First, on arrival, I sought directions from a man who spoke English and insisted first that he have a drink and then, together with the barman, made me feel I had to pay for it. How foolish I was and felt. Then I was tricked out of some pesetas by a student, or that is what he claimed to be, who promised me tickets to watch flamenco dancing. He failed to turn up at the appointed place to meet his gullible victim.

I made short expeditions to General Francisco Franco's palace and the royal monastery El Escorial, the largest Renaissance building in the world, and a most interesting day's visit to picturesque Toledo to see El Greco paintings. Their somber colors and gaunt, attenuated faces and hands from a distant world of ascetics were strangely and powerfully magnetic. El Greco's paintings made a big impression on me when I first saw some in London and even more so in their hometown. I could have stayed for days.

From Madrid I took the train to Granada, where the undoubted high-light was visiting the Alhambra. It was spectacular, whether seen at a distance or from within, and like nothing else I had seen. Many years later Rosemary and I visited the Alhambra at a time when it was difficult to get tickets, but on my first visit I just walked in, as did everyone else. The Moorish influence was so palpable, I felt that Morocco itself could not be far away. Another excitement was the flamenco singing and danc-ing in a Gypsy (Roma) cave in the Albaicín quarter. My tour of southern Spain ended in Seville, another fascinating place of mixed Moorish and Spanish culture. I tried to go to the nature reserve at Coto Doñana, failed for some reason, and instead took a day trip to Huelva, on the coast, for a day's bird-watching. I was getting better at Spanish; I relished the sun and heat, the food, the buildings, the role of being a solitary exploring tourist-cum-naturalist; and I was sad to leave. I shared a train compartment on a long nighttime train journey from Madrid to Lon-don with a young woman. Pleasant conversation from diagonally op-posite corners, but understandably guarded; she needed no help in London, *gracias*.

5

Becoming Serious

I WENT up to Cambridge by train with my treasured bicycle, a pack, a suitcase, and a hockey stick, destination Selwyn College (Fig. 5.1), supported once again by a grant from the Croydon County Council. The first year was very exciting, as an entirely new world opened up of college life, lectures and labs, and extramural activities. I took zoology, botany, and organic chemistry courses in the first year; zoology, botany, and geology in the second for Tripos part one (an exam); and zoology for part two. I loved it all. I also loved sport, and combining work with play was a most enjoyable return to, and extension of, school life, but I nearly came a cropper by giving each of them equal time and attention.

From the perspective of my future career, day one at university was pivotal. I thought I would concentrate on botany because I knew more than I did in zoology. The first lecture I attended on the first Monday morning was in the Botany Department. It was given by Dr. Evans, a pompous individual who talked about drinking sherry with famous people he knew. The next lecture was in zoology and was given by Professor Carl Pantin. It could not have been more different: exciting, delivered at just the right speed, nicely illustrated, and full of fascinating questions and discoveries about invertebrate animals, their diversity, and how they overcame problems of living on land or in water. That did it: I switched allegiance there and then. The excitement continued with beautifully clear lectures on insect physiology by Professor Vincent Wigglesworth. Like Carl Pantin, he had enviable skill in telling a story by articulating a problem—how could an insect do this?—and gradually

FIGURE 5.1. **Upper:** At the University of Cambridge, UK, November 1959. **Lower:** Selwyn College, Cambridge.

revealing an answer through clever experiments. It was biological detective work and very stimulating. Organic chemistry had no intrinsic appeal for me, but the lecturer, Professor Saunders, did a creditable job of making chemical compounds and interactions a natural way of looking at everyday things like butter and milk.

Moving on to the second year, I greatly enjoyed the vertebrate half of zoology, given by Francis Parrington with a large component of paleontology. Foremost among other subjects with strong appeal were ecology, taught by George Salt, and animal behavior, by Robert Hinde and Bill Thorpe, because they were closest to my naturalist interests. Instruction was as much hands-on as possible. In lab practicals we learned the internal anatomy of snails and crayfish, the reproductive system of the rabbit, and the nervous system of the skate. This was classical zoology: find the internal anatomy, identify it, and draw it. I still have the handout sheets and my drawings.

We left the lab to learn field ecology on Sheep's Green and animal (bird) behavior out at the Madingley field station. Field trips to local gravel pits, clay pits, and rocky exposures of fossiliferous strata made geology real and tangible, a way of uniting studies of the environment past and present. Paleontology was especially exciting because the footprint of gradual evolutionary change emerged when we compared fossil oyster shells and sea urchins in chronological sequences from the dated geological strata. A geological week on the Scottish island of Arran showed us the influence of more recent environmental change as reflected in raised beaches. I would have gone on a marine biology field course in the first summer if I had not done a course already at Plymouth when I was at school, and anyway I had plans to be on the Continent.

My friends were fellow bird-watchers and contemporaries Pat Bateson, Fred Cooke, and Hilary Fry, and also Bill Fry, who had a passion for pycnogonids, strange marine arthropods commonly known as sea spiders. The future naturalist and population geneticist Bill Hamilton was in the same class, and although I knew him, we were not friends. Girlfriends I did not have. I joined the Cambridge Natural History Society and Bird Club. On one memorable winter's morning I nearly froze in coastal Norfolk when helping to take birds out of mist nets for others to ring (band). This was my first experience of mist netting. We were almost overwhelmed with greenfinches and linnets.

Right from the beginning, I knew university was the environment for me, the same revelation I had experienced on entering Whitgift ten years earlier. A benefit of living in college was close proximity to others one might never have chosen as colleagues. John Morton (biochemistry, veterinary medicine) from Chard, Somerset; Mike Young (organic chemistry) from Skegness, Lincolnshire; and Philip Crowe (religion) from Wrexham, in Wales; and I formed a close-knit group. Geographically, socially, and academically, we were a highly diverse set and a stimulating mini-society as a result. Academically, what I instantly appreciated at Cambridge was being treated as an adult whose opinions were of interest to those who taught us. The closest interactions were in tutorials. My first was with David Harrison, a chemist, a kindly man with surprising modesty and almost shyness that incongruously accompanied an unambiguously strong intellect. He was a (new) fellow of Selwyn, but I also had tutorials in other colleges—for example, with Professor Charles Whittingham, an expert on every angle of photosynthesis one could possibly want to know and the author of *the* textbook on the subject.

I learned most from another Selwyn fellow, Hugh Cott. He was a Victorian naturalist, or so it seemed to me and the two other students in Selwyn studying biology in my cohort (three was the typical number of students in a tutorial). Cott had done pioneering work on crocodile behavior in Africa, on which he wrote a monograph, and on animal coloration. His book, *Adaptive Coloration in Animals*, full of fascinating natural history, became the authoritative work on animal visual deception. Additionally, he was an excellent photographer and artist. During World War II he applied his knowledge of principles of countershading and background matching to camouflaging the military equipment in the North Africa campaign.

Tutorials were fun because they taught us a lot in a friendly atmosphere, and the teachings were occasionally interspersed with lighthearted comments and anecdotes. Cott described catching small migrant birds and killing them for an experiment to test whether boldly marked plumage—for example, black and white—served as a warning to a would-be predator that it was distasteful, and less conspicuously marked plumage did not. He pegged out the carcasses on the Egyptian and Libyan desert floor, leaving them overnight and checking the following morning to see which ones had been eaten, most likely by hedgehogs, and which ones had been left. Ones from birds with conspicuous plumage "survived" best, as I recall.

Another anecdote was frightening. He was nearly caught behind his tripod and camera by a crocodile on its way from its nest to the river. With stories like that, he rose in our admiration. Yet another was as amazing as it was amusing. His Land Rover broke down with the second of two irreparable punctures in the middle of nowhere. After a long time, his African assistant went off to relieve himself—"doing his business" in Cott's quaint language. His helper noticed a peculiar circular growth in the grass around where he had just defecated. It turned out this was caused by a half-buried wheel, and when they excavated it, they discovered to their great joy and relief that it had a perfectly functional tire and could be fitted onto their vehicle. They drove off!

The university offered lecture series that were open to the public, and I was drawn to one in particular, a series of six lectures, or perhaps more, on Gothic architecture by the German historian Nikolaus Pevsner. They were fascinating and revelatory. In this period of my intellectual awakening, I could have imagined architecture as an alternative career, or archaeology and anthropology, if I had been forced out of biology and geology for some reason. Copiously illustrated, the lectures took us on a tour of the cathedrals in Europe, and I learned a new lexicon of *triforium*, *clerestory*, *transept*, and *apse*. Notre-Dame had impressed me on my Paris visit, and Canterbury Cathedral had done the same during national service, so already I had a good idea of the beauty and atmosphere of cathedrals. From Pevsner I learned how the atmosphere is created architecturally, and how an impression of lightness of the columns is achieved by fluting, and lightness of the ceiling vaults is achieved by subtleties of tracery (ribs) and decoration. It was another detective story, like the zoology lectures, with hidden meaning revealed.

One of Pevsner's examples was the cathedral at nearby Ely, originally Norman, with Gothic features added later. The cathedral was built in the eleventh century, and its most notable feature was an octagonal tower with a painted wooden lantern in the ceiling above it (Anderson and Hicks 1978). I had to see it, so one weekend I got on my bicycle and rode a distance of about fifteen miles to the Isle of Ely—or Isle of Eels as it used to be called. Viewed from the surrounding flatlands the building looked magnificent. Viewing it from the inside, I could appreciate how the ethereal effects were created. Experiencing these splendors was like a lab practical as complement to the Pevsner lectures. Ely has a special place in English history. Surrounded by fens (marshes) in the past, and difficult to approach and capture, Ely is a place for imagining what it was like to be living one or two thousand years ago. Safe from invaders (except the Danes) but probably malarial and not very healthy.

I took off as much time as possible to play hockey in the two winter terms and cricket in the summer. I made the greatest advance with hockey, captaining the college team and playing many matches with the Wanderers, the university's second team, even captaining them on one occasion. I came close to playing for the first team when the center half came down with the flu just before the match with Oxford, and the substitute called upon was either the left half or the right half of the Wanderers. I lost the competition to a better player. With cricket I started with the success of being included in the freshman trial. We had training in the nets under the watchful eyes of two national-team players, Tom Graveney and Ted Dexter. I hoped to secure a place in the university second team, but it never happened. However, a substantial fringe benefit of being secretary or captain of a college team was being able to stay in college and avoid seeking "digs" (accommodations) in the town in the second year. My room was one floor up on staircase B, the first staircase to the left of the main entrance, home for three extremely enjoyable years.

At the end of the first year, I put a pack on my back and headed for Brussels to attend the 1958 World's Fair. The exhibition was exciting for a young person with an international outlook because every country's pavilion, a showcase of its national specialties, was fascinatingly different, and also because new technologies were on display. The Czechoslovakian pavilion was a standout. It had a film projecting onto a moving glass screen that gave the impression of three-dimensional viewing. I stayed in a hostel with students from many countries and joined a small group, just as at Selwyn. One member was an Italian whose father was a judge; one was the son of a professor from Paris; and two others were from southern Spain. We were all university students at that delightful age of exploring the world with increasing confidence. The lingua franca was English. The Italian and French were intellectuals and could tell me more about Shakespeare and the kings and queens of England than I could tell them. The Spaniards were fun-loving free spirits dressed in long, flowing robes of eye-catching green and black, they went around with their guitars as troubadours, singing with that same vigor and vivacity as their soul mates at Figueras (chapter 4). All of us really bonded for the three days we were there: an unforgettable and happy experience that fed my desire for more international travel with social contact.

From Brussels I hitchhiked and walked to Ostend, then took the ferry to Copenhagen to be best man at Ron Wilson's wedding to a Danish woman, Birthe, daughter of a dentist and his wife. Ron had lived around the corner from us in Norbury. I felt very sorry for him when his father committed suicide, and that probably explains why I stuck with him even after we had a falling-out. After the wedding I took a ferry to the charming Danish island of Bornholm, stayed in a youth hostel, and walked all over the island. Roses were in full bloom and scent, the sun shone, and warblers sang in the hedgerows. It was a perfect time of the year. From there I took a ferry to southern Sweden, intending to visit Stockholm, but the Swedish driver who gave me a lift told me that if I had only a little money, I should stay in Norway because it was much cheaper. It was good advice. At a fork in the highway, he drove off to the right, leaving me to seek a ride to the left. I slept out that night, and found a tick in my thigh on waking, but then heard and saw my first Pied Flycatcher. I continued the hitchhiking journey to Oslo. This was easy to do because of a strong and widely shared pro-British sentiment. Several Norwegians who picked me up told me they were so grateful to the British for fighting on their behalf in World War II. I had taken some good advice and traveled with a Union Jack attached to my pack.

When I reached Oslo, I stayed in a splendid youth hostel, almost a hotel and nothing like the primitive English ones I had experienced. An enduring memory is walking into a large, steam-filled shower room and gradually becoming aware that most of the bodies I could dimly make out were much more like those of women than men—being unprepared and slightly scared I hastily retreated. At the hostel I learned that I could earn money by unloading bananas from a boat. I was short of money, so I went down to the docks and signed on for three days. We were warned that mambas had been known to crawl out, although that may have been apocryphal. The pay was good, and it gave me enough to continue, first to Odda, in the mountains, where I had a memorable meal of fresh herring at another lovely youth hostel and learned a lot about local geology from an American student doing field research, and then on to Bergen. The historically distinctive feature of Bergen is the waterfront of colorful houses. They are the descendants or replacements of the original houses that were built centuries ago by the Hanseatic League of German traders and merchants for storing fish and other products.

I made one additional excursion before leaving and heading back south, and this was to Edvard Grieg's dark-paneled house in the woods, a pilgrimage of sorts, since I loved Grieg's music. I walked down a birch-fringed lane in "English" drizzling rain, conjuring a mental picture of the inspiration from nature that the artist creatively translated into music. I returned home, and on arriving in London I had one shilling and sixpence in my pocket. I cannot be sure, but I think I took a bus home to Norbury. However, I do remember buying a Cadbury's milk chocolate bar for sixpence and devouring it. I am sure I had lost a lot of weight while on the road. Looking back, I wonder: Where was my safety margin? Nowhere!

That first summer vacation gave me an appetite for more travel, so in the next and final summer vacation I traveled to North America, which I describe in the next chapter. Returning home, I received the shocking news that I had been given a third grade, of passing, in the Tripos part one exams. My tutor at Selwyn, Philip Durrant, a chemist, advised me to think of studying in my last year for an ordinary degree and not for an honors degree (Tripos part two). He suggested teaching at a public school would be a good career to aim for because I could combine my evident love of and success in sports with teaching science, just as my careers master had said at school. That was my fallback idea all along, plan B, but not something I wanted to pursue as first choice, because I had become deeply interested in biology as a subject to explore. I pleaded that I would apply myself to work with greater effort and give myself more time to do so by reducing my sporting activities. My sincerity managed to persuade Dr. Durrant to allow me to continue with the honors program. Sporting indulgence had caught up with me and threatened to dictate my future, so I gave up the captaincy of the college cricket team.

In Tripos part two of the zoology degree program, we could either do a research project and attend a few lectures or take courses of just lectures. I was interested in everything biological at that time but too much of a generalist and not ready for semi-independent research at the required high level, so I took the lecture option and benefited enormously. Probably the most thrilling lectures were given by Sydney Brenner: four lectures on genetics, covering classical genetics in the first and molecular genetics in the following three. The lectures were brilliant, and so were the problem-solving stories he unfolded before us. I had known little of genetics at the time, and no biochemistry, but if I had been better prepared, this is a subject I could have easily chosen for later specialization. More than a half century later, I had the chance to thank Dr. Brenner (chapter 20).

I admired my colleagues who seemed to know exactly what they wanted to do in life after Cambridge. Like crabs, their carapaces had hardened into a variety of shapes appropriate for their future career niches, whereas I was still in the terminal molt without a commitment to a final growth form. Nonetheless, by the third and last year I was becoming more serious about biology as a basis for a career (plan A). For one thing, I developed a surprising interest and enjoyment in working in the quietness of the Newton Library of the Zoology Department, immersing myself in the scientific papers we were assigned to read and the books I took down from the shelves to browse. For another, I had become fascinated by some of the exhibits in the University Museum of Zoology. One was the Kakapo, a green, nocturnal, flightless, herbivorous owl-like parrot of New Zealand. This was an outrageous ecological misfit, I thought. I had learned at school about another one, the Galápagos Cormorant. It too was flightless, like other birds in isolated parts of the world. How had that happened, and why? Why does a bird give up its flight, and a bright green parrot give up the light, to go foraging on the ground in the middle of the night? These two unusual birds provoked deeper thoughts than my usual superficial ones about adaptation, although nothing profound; rather, they disturbed my complacency in uncritically accepting what I had been taught, and without my realizing it, they helped me to start questioning, probing, and thinking for myself—late, yes, but better late than never! Little things like these can have big effects.

I started thinking of my career and working toward a PhD degree. Given my ornithological interests, the obvious person to work with was David Lack at Oxford, but Pat Bateson warned me that he could be "a bit of a martinet"; furthermore, Lack was deeply involved in research on radar-tracking of migratory birds and not, at that time, interested as much as he had been in bird ecology and evolution. I had been stimulated by learning that single genes influenced traits such as spots on butterfly wings and bands on snail shells and thought there must be a way to investigate genetics of birds but had no idea how to do it. These tentative ideas seemed to lead nowhere.

I studied hard for the finals but not very efficiently and was awarded a 2:2—that is, in the lower half of a second-class degree. I like to think it was close to a 2:1 and, like my scholastic standing at Whitgift, about average. The practical exam was, for me, traumatic. We had to dissect, expose, draw, and label the parasympathetic nervous system of a skate, a type of cartilaginous fish belonging to the same order as the rays. I had just discovered that I had done everything wrong when I was called in to have an oral exam. My self-confidence and calmness had deserted me. I explained my predicament to the examiners, Drs. Ramsey and Parry, who I could see tried their level best to relax me. Finding my special interest was in birds, they asked whether I had read David Lack's little book *The Life of the Robin*. Yes, I said, I had. The conversation moved on to a question about an experiment in which Lack had presented territorial birds with a tuft of the robin's red breast feathers to elicit an aggressive response. Did I think it was a good experiment? Yes, I did. If so, then what did I think a robin would do when close to a field of poppies? I should have said, "Go beserk!" I had no answer then but plenty of answers later when I went over and over this embarrassing experience, picking at the scab of the wound.

My departure from Cambridge was much more enjoyable. I took Elizabeth, a fellow zoologist, to the Trinity College May Ball, and then we visited the celebrations at other colleges. How we stayed awake all night, I have no idea. The next morning, we traveled down to London by train, parted at King's Cross station, and from there I rode my bicycle home, with a pack on my back and a cricket bag with bat across the handlebars. I had not quite finished with Cambridge. The Selwyn cricket team went on tour to Holland, and we were shown how to play by some very competent Dutch players. Back at Cambridge I received my degree, witnessed by a proud father, Auntie Vi, and Uncle George.

6

Imprinting on Wild America

I JOINED an Oxford and Cambridge undergraduate travel society, not having the slightest suspicion that it would alter my life's trajectory. The society found vacation jobs for students in Canada and the United States and arranged a cheap charter flight to get us there and back. I took a Greyhound bus from New York to Toronto and then hitchhiked across Canada to Vancouver, where I had a job with Downtown Parking Corporation for two months. The journey across Canada was spectacular, even including the flat prairies beneath enormous skies and sometimes mountainous clouds. In Manitoba I was picked up by a young military man who smuggled me into his base camp for the night. I remember two things about the following morning: his extreme reluctance to get up early to drive me off the camp and to the road ("but you promised") and being woken up by the extraordinarily powerful and musical song of a meadowlark—as if someone was creating the sound of running water by playing a woodwind instrument in my right ear. Continuing westward, I visited Lake Louise and admired the grandeur of the place, then nearly froze that night when camping out. British Columbia made the strongest impression on me by virtue of the scale and wildness of the forests and the mountains and the wildlife. One driver took me over a mountain pass after dark, and we saw a lynx and a porcupine, such exotic animals living in wilderness. It was much wilder than anything in Britain.

I stayed a couple of nights with Ron Wilson and Birthe, who had moved from Denmark to Vancouver, then I changed to a small, single basement room in 6 Bute Street for a weekly rent of six dollars and fifty cents. I had to live cheaply, as my pay from Downtown Parking was only thirty-five dollars a week. People left their cars with the keys in the ignition before going to work, and my job was to park them. My first effort to do this, for an attractive young woman called Linda, was a disaster. Her car was a light green Morris Minor 1000. "No problem," thought I, as I had learned to drive on a Morris Minor in Norbury (but twice failed a driving test). I only had to turn the wheel and drive a few yards, but in that short distance I managed to scrape the vehicle against a neighbor. "Don't worry," I was told by my boss, Sam. "The company's insurance will cover it." My next experience was no better. Many of the cars were really beaten up—old rusty derelicts—and my next task was to park one of these. It was an enormous American Pontiac or Chevrolet, with an automatic transmission, which was new to me. All was well when I drove straight between two widely spaced rows of cars, until I reached the end and turned to the left. To accomplish this with power steering, one had to turn the steering wheel a few revolutions, whereas in my Morris Minor training I could do it with a quarter turn. And that's what I did but, unfortunately, I also kept going. The result was three more damaged vehicle bodies to add to the previous two. Actually, I don't think the owners ever knew. I was transferred laterally to the role of ticket collector.

I went up the Squamish Highway with Ron and Birthe on one day off, visited Stanley Park more than once, and paid a visit to the University of British Columbia (UBC). I managed to get an interview on the spot with Ian McTaggart Cowan, head of the Department of Zoology, who I had seen on television at home on a program about invertebrate life on the Vancouver-area seashore. Both he and the campus impressed me, and I left determined to apply for graduate work. I had saved money by living frugally, and it enabled me to finish my North American experience with a grand tour of the United States.

From Vancouver I hitchhiked south, then across the southern tier of states from California to Florida, back up north to Washington, DC, and thence went via Greyhound bus to New York City before returning to England. I had been inspired by reading *Wild America*, a book written by ornithologists James Fisher (UK) and Roger Tory Peterson (US) about their bird-watching extravaganza on a long drive around the States in 1953 (Peterson and Fisher 1955). My trip too was enormously enriching, for the great diversity of people I met, the scenery, and the national parks I visited.

For most of the trip, hitchhiking was easy, especially in the West. My first ride was with a very friendly young man named Howard who took me for a short distance beyond the US-Canadian border. I wrote down his name and address in my diary and told him I would send him a postcard when I got back home. This was to be repeated many times during my journey. Sadly, I lost the diary several weeks later, and failure to honor my promise has been a source of guilt ever since. On the first night out in Washington State, I slept in a little roadside hut with only three walls. I discovered I had floodlit a rural airport by turning on a switch that did not seem to work only when a police car screeched to a halt and out came two policemen. Actually, they were friendly when they found out how innocently inoffensive I was and let me stay there overnight. It rained in the night, but the hut had a roof.

I continued, and the next day had the good fortune to be taken to Crater Lake, incredibly blue next to the deep green coniferous forest. It looked starkly beautiful but cold and lifeless. In contrast, the aromatic air in the brilliant sunlight was exhilarating, and it felt good to be alive. From Oregon thence to the San Francisco Bay Area and a visit to the University of California at Berkeley. The campus impressed me, just as UBC had done, and I decided to apply for graduate school there too. I don't know where I stayed in places like San Francisco—probably YMCAs or youth hostels, cheaply, anyway—and I walked a lot. I visited Golden Gate Park and Fisherman's Wharf in San Francisco; some of the men looked English in their sports jackets. The sprinklers in the park never stopped, even in the rain.

Visiting Yosemite in early September was my second piece of traveler's luck. As at Crater Lake, the dry air, aromatic scents, bright light, and contrasting colors were almost intoxicating, and the views of the landscape from many points were spectacular. It was no wonder that John Muir had made Yosemite his home and worshipped Half Dome like a shrine. Returning to the lowlands, I continued my journey and somewhere, perhaps near Monterey, I met Floyd St. Clair, an American from the Romance languages department at Rutgers University, and David Watmough, an English artist—Floyd tall and thin, David short and stout, and both lively conversationalists. They took me to Point Lobos (Fig. 6.1) to watch Sea Otters feeding among the kelp beds and southward along the coast, stopping at interesting places like the Santa Barbara Mission, built by the successor of Father Junípero Serra. In Pasadena I stayed with Walter and Mary Evans, who I had befriended in Norway a year earlier ("If ever you come to LA, you must stay with us"). They took me to look at the University of California at Los Angeles (UCLA) and University of Southern California (USC) campuses, and I added them to my collection of potential graduate venues. USC especially looked splendid, and together with the diversity of habitats I had passed through, the campus increased my desire to move west for graduate work. I visited La Brea Tar Pits, famous for the exotic mammals that had been trapped and died thousands of years ago. I had learned about saber-toothed tigers at Cambridge, and here they were, like an extended laboratory experience making real what previously had been only a photograph and a drawing in a book.

I left Los Angeles after a couple of days, aiming for New Orleans via Arizona, New Mexico, and Texas. This part of the odyssey was the most thrilling to the budding naturalist: wonderful scenery, entirely new habitats including deserts, and strange new animals like Kaibab Squirrels and Greater Roadrunners. I was enthralled. The highlights were Death Valley, the Grand Canyon, Meteor Crater, and the Painted Desert. In Texas I loved the juniper-grassland habitat. The fact that I was able to experience this extraordinary variety was due to the third piece of luck. I was picked up by Andy Anderson, an American military man, and his English wife, Doreen, and they drove me all the way from the Grand Canyon to the outskirts of Dallas–Fort Worth, a distance of nearly twelve hundred miles. That Union Jack on my pack was a golden passport! When they stopped for the night in a hotel, I slept out or in a much cheaper place; they picked me up the next day, and the journey resumed, sometimes after I had used their shower. I think they enjoyed

FIGURE 6.1. With David Watmough, at Point Lobos, California, September 7, 1959.

the national parks as much as I did. I was impressed, not only by the parks themselves and what they contained and protected but also by the museums, displays, and information. I had never experienced a national park before this North American trip.

In southern Texas I was picked up by a completely different kind of family. They were driving an old beat-up American car, a Chevrolet or Ford, I would guess, and I was squeezed into the back with two or three children. The parents told me they were the poor whites of the country. Their car and their clothes said the same thing, as did the unkempt hair and grubby faces of the children. In contrast to their appearance was their warmth, generosity, lively intelligence, and courteous friendliness. They had every reason to complain about their circumstances, but instead they were a happy family. My heart went out to them, and I can say, this family moved me like no other I encountered. I declined their offer to stay with them in their shack, and told them, yes, I would definitely send them a postcard when I reached the end of my travels to tell them how I got on. But, as I have mentioned above, I failed to do this, as I lost my address book. After receiving their generosity, I felt dreadful; I still do and will continue to do so to my grave, at the thought of letting them down and not being the responsible young Englishman they thought I was. Years later I found my address book and wondered whether I had misremembered not writing to them. I hope so, but this is hardly likely. The address book shows an entry for Bill Caskey, Route 2, Box 435-S, Conroe, Texas.

My approach to New Orleans was as unpredictable as any part of the journey: I drove there. A woman picked me up so that she could be driven, because she was too tired to go on. I protested that I did not have a valid driving license. "That doesn't matter," she said. "Nobody will check you." I passively but nervously accepted my fate and got into the driver's seat, recalling what happened when I last drove a car with an automatic transmission, and off we went. She explained that she was getting married in New Orleans, and this was marriage number three. She was about fifty. After quite a while we got a flat tire—not surprising, given the state of the rest of the car. We got out, and she just assumed I would fix it. Instead, she had to direct me to lift the car up with a jack, take the wheel off, and put the spare one on. Mercifully, it worked, and we completed the journey. My part ended at a tangled mixture of roads and overpasses just as it was getting dark. She warned me that it could be quite dangerous to sleep out because of antagonism between blacks and whites, and with that she left. I found a secluded place beneath the overpass and had no scares or difficulties, but I was anxious.

The next day I got a ride into town and checked in to a YMCA. That day was eventful, as I was accosted by police no fewer than three times. The first time I didn't have my passport, and they drove me to the YMCA to prove who I was. The third time, I had the courage to ask the policeman why I was repeatedly apprehended, and he explained that a young Dutchman who looked like me was on the run, suspected of having murdered a stewardess on a liner that had docked in Boston a couple of days earlier. Later they caught him, and I saw his photograph in a newspaper, and sure enough he did look like me. He was twenty-nine though, and I was twenty-two. I enjoyed the rest of a brief stay in New Orleans, visiting Bourbon Street and seeing a European version of a US city.

The next segment of the journey was in the Deep South, heading east toward Florida. I had seen pictures of magnificent mansions, white, colonnaded, and festooned with Spanish moss, and so it was a minor thrill to see them from the car. I was now traveling in a biologically rich area, fascinated by what I could see from the car but regretting I had little opportunity to go walking or bird-watching. A particularly memorable ride was with the Oxford-trained Reverend O. C. Edwards Jr., a theological teacher. He invited me to his house and gave me a fine dinner and a bed for the night. I recollect that his home was in Morgan City, Louisiana. He explained the history of this part of the country, race relations, and the controversial policies of people like Huey P. Long. Once, when I was riding in a bus somewhere in Alabama, the conductor pointed to a sign that said, "Whites Only." I was sitting in the Blacks-only section. "That's OK," I said, not realizing I was being innocently provocative. The penny dropped later. Black people just stared at me. Segregation on buses had been outlawed three years earlier by the Supreme Court, so I suppose this was a lingering attempt to maintain de facto segregation.

I arrived in Miami late one afternoon, bought a coffee and some bread, and started walking northward. I must have been starving, as I remember staring at the prices of fruit and buying a single apple or orange. I had walked toward Fort Lauderdale, and by now it was nighttime; however, there was plenty of light to find a comfortable hollow in the sand beneath casuarina trees, where I spent the night. Thankfully, no one was around, and I slept deeply. The next morning, I packed up and walked across the road to get breakfast at a café. I was greeted with the words, "Lucky we didn't get the hurricane." "Hurricane?!" "Yes, it just missed us, and it was a big one." Then I knew why the beach had been deserted. Only a naive Englishman would tempt providence by sleeping out on the beach before the arrival of a hurricane!

My main goal in Miami was to see if I could visit the Everglades, a landscape described so evocatively in *Wild America*. I walked back until I found the local office of the Audubon Society and inquired there. I was in luck, for the fourth time. They put me in touch with an enthusiastic volunteer named Dick Cunningham, who offered to take me in his car. Dick was an expert guide who could answer all my questions and take me to many of the good places. This was a thrilling day, largely making up for all the nature walks I did not take in the swamps and woodlands of Louisiana, Mississippi, and Alabama. The day was marked by new experiences—for example, seeing alligators sprawled across the road; walking among tropical trees in an attractive hammock, a slightly elevated habitat island, on the Anhinga Trail; and seeing strange birds, like the Anhinga and Limpkin as well as numerous herons. I took a photo of a green lizard and when I next looked, it had turned from green to brown!

Dick showed me a Water Moccasin (or Cottonmouth) swimming close to the bank where we were standing. On the long way back to Miami, he described how a snake enthusiast friend had managed to grab a Water Moccasin, with the intention of taking it back to the museum where he was working. He had a good grip of the snake behind its neck with one hand as he drove back into town with the other. However, the snake's coils put great pressure on his wrist and squeezed and pushed until it worked its way loose enough to bite him. Undeterred, the herpetologist changed plans and headed to the hospital without relinquishing the snake, and managed to arrive and hand over the snake, somehow, before collapsing. He survived. The famous scientist Edward O. Wilson relates a similar boyhood experience with an exceptionally long Cottonmouth (Wilson 1994, 90). In imminent danger of being being bitten by the snake, probably fatally, Ed had to hurl the snake into the bushes. I have never had the urge to handle a venomous snake.

I could have explored more, but Dick could spare only a day. He linked me up with a friend who was leaving for Virginia the next day. Herb Kinsey was a fine companion, about my age, and full of enthusiasm. On one occasion in North Carolina, we stopped at a small farm surrounded by corn and tobacco so I could photograph a typical homestead. The man came out, and Herb started a friendly conversation. I was entranced by the man's rural Southern accent. "You're not one of them revenooers, are yer?" he asked Herb. That means somebody from the Internal Revenue Service, explained Herb, wanting to tax the illicit alcohol that this man surely had.

I knew nothing of Herb's family's wealth, not even a hint of it, until we entered the driveway and I saw the magnificent mansion surrounded by tall trees. It was a smaller version of what I had seen in the Deep South. I did not see his father; I think he was a banker. My first impression of Herb's mother was, she is too young to be his mother. Dick and I went into separate bedrooms and had showers; Dick changed clothes and dressed for dinner, and perhaps I changed a shirt—I hope so. And then we descended to enter a magnificent dining room and a table prepared as if for some honored guest. Mother had changed into a dazzling white evening dress complete with sparkling jewelry, and she herself was pretty dazzling too. This was the only time I drank wine on this US tour. It was a memorable evening.

From there I headed to Washington, DC, where I decided to spend some of my little money on a Greyhound bus ticket, realizing how difficult it might be to get rides from there to New York. I liked DC but left after only a day of sightseeing, and I reached New York with very little money to spare, having spent eighty-nine dollars to travel around the United States in five to six weeks. This was an accomplishment, but I was also aware that the help I had been given in travel and occasionally food and accommodation, amounted to very much more. I tried to reciprocate in the only way possible, by contributing interesting conversation to our ephemeral friendships. It was a great experience for me, and I hope rewarding for my generous companions.

In New York City I stayed in a YMCA and rejoined some fellow Oxbridge travelers. The next day a couple of us decided to split up and be tourists in different places. I wanted to visit the Bowery, as I had read about its poverty and social problems, and also the United Nations, both of which I did. Listening to speeches in the UN was particularly exciting, and as witnesses to globally important issues, my companions and I felt optimistic about the future security of the world because it resided in this body. Before that excursion, and as we were trying to decide where to reconvene, I spied the sign "Restroom" and suggested it would be a good place to meet. We duly met there, but it wasn't the room of relaxed armchairs and sofas I had imagined it to be. I can still remember the shock and disappointment of entering that particular room of rest (a bathroom). It was my last piece of naivety in the United States. I managed to find my way to Idlewild

Airport (later renamed John F. Kennedy International), onto the correct plane, and so back to England.

Soon after returning I applied for graduate work at all the major universities on the US West Coast, plus Harvard and Yale. As a backup plan, I applied for a position at the Field Studies Council at Slapton Ley in Dorset, England. I was interviewed there and thought it would be a nice place to get some experience before doing something else, but it came to nothing. They were looking for and found somebody much more experienced with insects than I was. So, when I was graciously offered a place at UBC by Ian McTaggart Cowan, whom I had met in Vancouver, I did not wait for news of the others but accepted the offer straightaway. Perhaps my interview with him was a key factor. To help finance it, I was strongly advised to earn a hundred pounds over the summer, which I did locally at Mitcham, near my home, by working for Scaffolding Great Britain (SGB) for two months. My job was to oil rusty scaffolding clamps, loosen the rust, separate the parts, and throw each one into its prescribed bin, and again, all day. As dull as anything, and not nearly as interesting as my winter odd jobs of being a temporary postman at Christmastime and working as a checkout person at Sainsbury's grocery (later called a supermarket). But I stuck to it, as I had a goal to work toward.

Before setting off to the New World, to a new life, and to the discovery that combining teaching with research was the career for me, I made one more foray to continental Europe, with Mother, her husband, Bill Baker, and my half sisters, Sarah and Philippa (Phip). I had bonded well with Bill by helping him with odd jobs in the vacations, painting on the fascia boards and signs of shops. We set off for Spain in a van and got as far as Málaga before time ran out and we had to return. We passed through the flat, central, grape-growing area of the country that I had visited in 1957. I enjoyed the trip, the sun, and the opportunity to use my limited Spanish, but my sensors were not attuned to the tensions in the family that made it less enjoyable for the others.

7

An Island Problem

A WORKERS' STRIKE prevented me from sailing from Liverpool to Montreal, so I flew, and then I hitchhiked across Canada. Returning to Vancouver and to UBC was exciting. I shared a basement room just off campus in West Sixteenth Avenue with Keith, an Englishman, and Murray, a Canadian. On my first day in the department, I met Mary Jackson, who ran the labs for Zoology 101, the introductory course for which I was to be a paid teaching assistant. I also met Bristol Foster, my office mate. He was a year ahead of me, had traveled for a year in Africa before returning for graduate work, and was, as I soon learned, an excellent naturalist. That was Friday, and on Monday I met my wife to be. My brain contains a video clip of the momentous occasion. It runs as follows.

I emerge from my office on the fourth floor holding a pair of skis, and coming down the corridor from the left is Mary Jackson with a companion. "Peter," Mary says when they draw level, "I would like you to meet another English person. This is Rosemary Matchett, and Rosemary, this is Peter Grant" (Fig. 7.1). We shake hands. Then the first words: "Those skis are too small for you," Rosemary says. "No, they are not," I reply, equally firmly. "Yes, they are, I can see just by looking at them." "No, they are not, and I know that because they were sold to me by my Canadian landlady, and she should know"—my reply, I am sure, is a winner. And in a strategic lateral move in the conversation that I have come to know so well, Rosemary says, "The way you tell is to lift up your arm, bend your hand over and the tip of the ski should fit just inside the bend of the wrist." "All right," I reply, fully confident of the outcome, and

FIGURE 7.1. Rosemary Matchett—the girl who bought my skis—at the University of British Columbia, 1961.

do so, only to find that my arm is a good four inches too long, or rather the skis are four inches too short. "You are right," and I think that is the end of it, but no—Rosemary continues: "They would fit me perfectly." Already on the back foot, I hand them over, and sure enough they fit perfectly by the bent-wrist test. "I will buy them from you. How much did you pay for them?" Rosemary says. "Eight dollars," I reply. Holding the skis, she hands me the eight dollars, we say goodbye to each other with somewhat less than mutual enthusiasm, and I retreat to my office.

And there are the poles! I had paid eight dollars for a pair of skis and poles and had sold the skis for the price of the total. This petty triumph did not last very long, however, because when the skiing season arrived, we were developing a boyfriend-girlfriend relationship, and I bought her a new pair of ski poles for fifty dollars. At least that is how I want to remember the price, but it was probably closer to fifteen dollars.

The second most important thing in that first year was embarking on research. For a supervisor I had a choice between Jim Bendell and Miklos Udvardy. Jim studied Blue Grouse population dynamics, and Miklos (Nick) studied biogeographic patterns of bird distributions. The science would have been better with big genial Jim, but his group of students were too closely connected to his main interest, and there seemed little room for independent growth. It seemed too much like a grown-up group of well-bonded Boy Scouts. Udvardy, on the other hand, was a cosmopolitan conversationalist, a European generalist scholar, and a Hungarian who had escaped from the communists ahead of their invasion of the country. I chose him. Frank Tompa, also a Hungarian refugee, joined Nick's group as a graduate student at the same time.

Next was the question of a research topic. My main interest was in competition between bird species that came from reading David Lack's stimulating small book on Darwin's finches in the Galápagos archipelago (Lack 1947) while I was at Cambridge. I decided to pursue this subject in greater depth. I dithered about, thinking uncreatively of studying competitive interactions between a pair of ecologically similar species, such as Black-capped and Rufous-backed Chickadees, or Cinnamon and Blue-winged Teal. I was rescued by a stroke of good fortune that set me on a trajectory I have held to for the whole of my career: a chance to work on an uninhabited island.

The opportunity came out of the blue. H. R. MacMillan, head of the MacMillan, Bloedel and Powell River logging company, had approached the fisheries part of the UBC Department of Zoology with a request for a fish expert to join him on his regular visits to Mexico on his yacht (almost a liner) and to identify all the fish he caught—marlin, billfishes, and smaller ones. The waters around the Tres Marías off the west coast of Mexico are particularly rich in marine life. Peter Larkin, our superb fisheries biologist and statistician, went on two trips, accompanied by Ian McTaggart Cowan, and while there they went ashore and collected specimens of birds for the department's museum. When I joined the department, Ian persuaded H.R. to fund a graduate student's research. I was the lucky one to be chosen, and a golden opportunity was handed to me on a plate.

"How would you like to study the birds of the Tres Marías islands for a PhD?" I was asked. I accepted the proposal without much reflection. From this distance in time, it seems like a leap into the complete unknown and far, far, far removed from the problem-driven approach to choosing a research topic that I have encouraged in my own students. Nevertheless, and with David Lack's book on Darwin's finches in my mind, I could see the potential of comparing island and adjacent mainland birds to see whether isolation on islands and the hypothetically relative freedom from competitors and predators had systematically affected the evolution of the island birds. This would be original, as I knew of no such study. Although I was in a master's degree program, placed there automatically in my first year, I was about to be transferred to the PhD program without, I should add, more than a mere formality of supervision.

Rosemary had studied genetics at the University of Edinburgh and was appointed as a lecturer and research associate in embryology, cytology, and genetics in the UBC Department of Zoology for just one year (B. R. Grant MS). She intended to return to Edinburgh for PhD studies. We discovered the joy of exchanging experiences, discussing art and biology, finding the many points in common in our respective views—but also the differences stemming from our different backgrounds—and simply enjoying being in each other's company on long walks in the mountains. Months later we bonded more closely on a week's skiing holiday at Banff. People tell you they "fell" in love. Well, I slid. I was cautious because I was aware my parents and one aunt and uncle had divorced. Eventually, it dawned on me: I could think of nothing other than Rosemary and decided I must be in love with her. We became engaged in April 1961. At Rosemary's request I wrote to her father and asked for the hand of his daughter in marriage. She explained he was very old-fashioned in some respects—he was born in 1903—and would greatly appreciate it. The reply was very amusing. He graciously granted my wish in formal language, then regretted that "there would be no dowry as her education has been expensively neglected."

We were married on January 4, 1962, in Saint James's Church in her home village of Arnside in Cumbria (Westmorland). The air was cold, clear, and frosty. We felt the village owned us for the day. From baker to caterers to those who decorated the church, it was their event, and all we had to do was to play our roles. I added an unscripted part. As Rosemary joined me at the altar and before the priest arrived, I whispered the top newspaper item of the day: "The DNA code has been cracked!" And so, we were married, and it took all of seven minutes by my estimation before we were pronounced husband and wife. I promised to love, honor, and cherish her—and have done so ever since—and she promised to love, honor, and obey. Obey? Rosemary? Our private contract, devised by Rosemary, was that she would obey everything she agreed to and no more. After a splendid reception (Fig. 7.2), we were whisked away in a Rolls-Royce and across the estuary for an overnight honeymoon at Cartmel, incongruously in the lea of a seventh-century priory. Our second honeymoon was a week spent in the daytime at the American Museum of Natural History in New York, measuring bird specimens for my thesis, and otherwise at restaurants and the Excelsior Hotel, near Central Park.

In Vancouver we lived just off campus on West Thirteenth Avenue (Fig. 7.3), and we bought a Volkswagen Beetle, a gift from Rosemary's parents. Life on campus was very enjoyable, but best of all were the field trips we made in that VW to different parts of the province—for example, a camping trip high up above the Ashnola River in spring to watch Bighorn Sheep, and another to the aspen parkland and potholes (glacial excavations) of the Chilcotin region in the fall to look at the migratory waterbirds. We visited Frank Tompa on Mandarte Island to see his study of Song Sparrows, toured the southern part of Vancouver Island, and hiked to Mount Garibaldi and Diamond Head to look at the alpine meadows.

On one of our explorations, we camped in a beautiful, forested location by Nicola Lake, with not a soul in sight. The name Nicola stayed in our minds, and we saved it for our first daughter. In the winter before we were married, we skied to Diamond Head, with sealskin strapped to

FIGURE 7.2. **Upper**: At our wedding celebration in Arnside, UK, January 4, 1962.
Lower: Estuary from Arnside village.

the bottoms of our skis to prevent slipping back, and stayed in a rusti-
cally comfortable chalet run by a Norwegian couple. Matronlike Millie
asked us if we were married, and when we said, no, not yet, she pointed
us firmly to bedrooms in opposite directions. That trip was memorable
because I broke both of my skis when I ran into a snowbank on what

FIGURE 7.3. Our first residence, at West Thirteenth Avenue and
Sasamat Street, Vancouver, BC, October 1961.

was intended to be the last run and had to borrow another pair to ski
out. In the final winter we lived in Vancouver we made a trip to Manning
Park with Mary Jackson and her boyfriend, George. Determined to
make an igloo, George had brought a book that told us how to do it. He
cut blocks of hardened snow with a machete, and we built the walls and
roof under his supervision. Then we slept in it. It was actually quite
snug. Our breath condensed and froze on the ceiling, giving it a smooth,
glassy surface. Sleeping in an igloo, hikes in alpine meadows, research
in the tropics—not surprisingly, we were in love with life as well as
with each other.

Intellectually I came to life in independent research and rapidly gained
competence and confidence after a slow start. In retrospect, I was unusually slow in converting a hobby into a career, owing partly to a conflict I had in Cambridge. On the one hand was my love of nature, which
crystallized into a fascination with the interplay of ecology, behavior,
genetics, and evolution. On the other hand was the education I had
received, which dazzled me with the power of the experimental method
to reveal with exquisite precision the physiological and biomechanical
mechanisms that animals, mainly insects, employed. However, the

emphasis on how organisms function left little room for the study of evolution in nature. I entered a different environment at UBC, one that encouraged fieldwork. The Department of Zoology was not then the powerhouse it has since become, but it gave me room to grow in a direction of my choice.

My first field season was a time for trying things out, making some false starts, being intellectually somewhat adrift, and having no supervision. Nevertheless, at the same time, knowing I was adrift forced me to think for myself and become a problem-solving scientist. To use a popular expression, I was on a steep learning curve, and I was training myself. The first question I chose to answer was whether the island birds differed from the mainland birds in size and proportions. As described below, the answer was yes. To explain the differences, I sought clues in the environment. My somewhat open-ended plan was to compare the ecology of island birds with their mainland counterparts in similar habitat. This would enable me to see whether differences in morphology or behavior could be attributed to the relative paucity of interacting species on the islands. Fewer potentially competitive species should mean greater scope for exploiting the environment, yet fewer predators should mean higher densities and, possibly as a consequence, competition for food. In short, the ecological niches of the species were either broader or narrower on the islands, and research was needed to find out which tendency prevailed. The plan of research was simple, but it was difficult to execute because the islands were not easily reached, so I began on the mainland. On most mornings throughout four summer months, I observed the feeding of bird species that were also present on the Tres Marías, counted what they fed on, and recorded how they did it. Also, I censused them.

The mainland location was Puerto Vallarta, in the Mexican state of Jalisco (Fig. 7.4). This was 1961, three years before *The Night of the Iguana* was filmed there with Ava Gardner and Richard Burton, when the sleepy little village, hardly a town, was transformed into a luxury tourist hot spot. I had spent my first year at UBC taking Introductory Spanish 101 in the new Buchanan Building, and now, in my first month at Puerto Vallarta, I was accompanied by an energetic graduate student, Arturo

FIGURE 7.4. At Hotel Rosita, Puerto Vallarta, Mexico, 1962.

Jiménez Guzmán, from the National Autonomous University of Mexico (UNAM) in Mexico City, to help improve it. The only problem was that Arturo insisted on practicing his English, so I learned much more Spanish after he left by having to speak it.

Rosemary flew down to join me for a few weeks and help with fieldwork, before she flew on to England to help her mother prepare for our forthcoming wedding. For lunch we bought a couple of roasted fish on the beach and combined this with slices of pineapple. The fish were cooked impaled on sticks inclined at forty-five degrees above flames rising from burning palm-nut husks. It was delicious and very cheap, and I could say the same about the aromatic corn tortillas we could hear being made early in the morning as the dough was slapped into a flat shape. There was one mini-market in town run by Ray, a friendly young American, and his Mexican wife, popular because of its ice cream.

After Rosemary left, I befriended Bob, a tobacco grower from Rhodesia staying, as I was, at the Hotel Rosita. One night at supper in the hotel a big, corpulent chief of police with a flamboyant mustache came in to celebrate his birthday with friends and family and proceeded to get increasingly loud and inebriated. He came over, sat down at our table, and demanded to know who we were. We explained. He started singing a Spanish song, telling us we had to join in. Then he pulled a pistol out of his holster and, pointing it at us, ordered us to sing an English song. Unperturbed, despite a gun wavering in front of us, I launched into a throaty rendering of "On Ilkley Moor bar t'hat" (On Ilkley Moor without a hat). That seemed to please him, and to celebrate he pointed his gun at the palm foliage above and fired. And then came the moment of truth: from quite some height a coconut fell!

After a long time of waiting, I received permission to go to the islands. The only boat I could get to take me there on that first trip, in August, the hurricane season, was a dugout canoe. The two fishermen and I made it to María Magdalena Island, a journey of one hundred miles, in one long day (Fig. 7.5). The island is a deciduous woodland with semidesert coastal vegetation. Beneath tall trees the ground was covered with leaves, a sight reminiscent of a northern beech forest. The birds were approachable and easily observed, as expected on an uninhabited island. I was thrilled by the experience of being alone on an uninhabited island where everything was new, and I threw myself into fieldwork for the few days we were there.

One day I decided to hike inland to reach high ground to look for the Brown-backed Solitaire, a famously melodious species of thrush. It was just too far, and I lost my way coming back but managed to keep calm, reasoning where I was likely to be, and as light was fading, I reached the beach, one arroyo (gulley) east of the one I had climbed up. Imagine the relief of eventually seeing the light of the fishermen's camp after a very long walk westward. When I reached camp, they told me they had come to look for me but could not find me and gave up. They handed me, in my dehydrated state, a cup of coffee that kept me awake for half the night, during which I solved the problem of competition with extraordinary power, clarity, and creativity as I winked in and out of sleep, only to find when I woke up that I had been repeating at intervals what I had been thinking for months.

On the way back, we broke down and completed the journey to the mainland near San Blas with a tattered sail that was more holes than

FIGURE 7.5. **Upper**: Beach camp on María Magdalena, Tres Marías islands, Mexico, August 1961. **Middle**: Arroyo on María Magdalena, 1961. **Lower**: On one of the Tres Marietas islands, Mexico, with a Brown Booby (*Sula leucogaster*), 1961.

cloth. We camped out in the lea of mangroves that night, and one of the fishermen was stung by a scorpion. We spent the whole of the next day under sail, limping back down the coast to Puerto Vallarta.

For the second and third field seasons, Rosemary and I drove down from Vancouver to Tepic, in Nayarit, and from there drove or flew to Puerto Vallarta. Getting from there to the Tres Marías was a headache because of threatening weather. An American, Travis Flippen (Flip), ran a business for sport fishermen, and I managed to get his help, twice in the second year and once more in the final year. His boat the *Bonito* was captained by Julian, a wonderful, ebullient man of no education but extraordinarily wise, competent, and resourceful.

Once we had just arrived at María Magdalena when a storm arose, so we transferred to a better anchorage at the neighboring island of María Cleofas. A boat came in, hailed us, and the captain pleaded with us to tow it back to port because the storm had broken the propeller shaft. Of course, we had to. So, on the understanding that he would pay Flip for the costs of fuel, we towed him back the next day. He was a sports fisherman from Hawaii, and as we later learned, he disappeared and never did pay.

On another occasion I went out with Nick Udvardy. When we returned, I found that Rosemary had barricaded herself into her room with furniture to keep out a very aggressive, squint-eyed, mentally challenged, and actually dangerous hotel odd-jobs man by the name of Pancho, who had tried but failed to break the door down with a machete. I arrived with a bad ear infection, no doubt from scratching an ear with dirty fingernails. With help from Señor Salvador, the well-named patrician hotel owner with a comfortable paunch, I found an excellent physician, Dr. Sahgún, who injected me with penicillin. This infection, plus firing a gun during national service, contributed to poor hearing in later years. I also inherited the wrong genes from my mother.

Rosemary and I went back to the islands again later. Before returning to work on María Magdalena, we reported to the civil and military authorities on María Madre, as required by the permit. There was an open penal colony on the island. We presented our papers to the *commandante*, and while waiting for them to be returned to us, we were approached by an extremely courteous official. In perfect English he asked whether we could give him books, any books, because he was starved of good literature. He looked the bookish sort. Just before leaving he explained he did not work for the government—he was a prisoner. We did not ask him what his crime had been—"cooking the books," perhaps.

Having experienced the woodland habitat on the island in my first year, I realized the area around Tepic in the highlands provided better, more similar habitat than the coastal Puerto Vallarta location. The Tepic-area woods were a mixture of oak and hornbeam before grading into pine forest at higher elevation. I chose a study area about five miles south and west of Tepic on the Jalcocotán road. Each morning we drove there, left the VW just off the road, and hiked up a trail past charcoal burners and heaps of smouldering charcoal. We were left alone to enjoy the leafy woods and do our work. Among the inhabitants was the Brown-backed Solitaire that I had missed seeing on María Magdalena. A popular cage bird, it has a beautiful, slightly melancholy song (https://ebird.org /species/brbsol1).

In Tepic we rented an apartment from Lewis Yaeger, a horticultural rancher from Texas, and his Mexican wife, Michaela (Micky). Lewis gave me a lot of help with finding particular birds, and I collected some for the UBC museum. However, this was a stressful time in several ways. Once, the gas heater in the kitchen blew up, almost in Rosemary's face. Rosemary went to the market alone to buy vegetables cheaply, because we were running out of money, but that was fraught with difficulties such as unwanted attention from the policemen. Nick Udvardy was failing to send any money despite repeated requests because, unknown to us, he was out of touch on extended vacation in Hawaii with his family. Finally, I sent a telegram with a plea for help to Ian Cowan, who immediately exploded into action, wired some money to a bank, and we were saved.

Undernourished but now solvent, we transferred to the Motel Cora, which was far better, and we stayed there for the rest of the time. It was situated next to highway 15, below a pine parkland on the edge of town, and was named after one of two indigenous groups that lived in the mountains. The other group was the Huichol. Occasionally we would see a few of the men in town, dressed in calico and wearing sandals and hats with tassels or dried hummingbird skins around the rim. They had come to sell their pottery and cloth handicrafts and get some supplies.

The motel was run by a nice, competent, and friendly woman, and we felt comfortable and secure. The rooms, on two floors, surrounded a large, cobbled patio. A young teenage girl went back and forth to help cleaning rooms, singing, and washing and sweeping the courtyard. A little boy called Enrique made himself a cheeky nuisance, and every now and then she would call out to him, "Enrique, venga!" (Enrique, come!). Unfortunately for the boy, a yellow-headed parrot, an Amazon with clipped wings, proved to be an excellent mimic and tormented poor Enrique with the same demand. Enrique would go rushing over to the girl, discover it was unnecessary, and walk all the way back to where he was sweeping, only to be fruitlessly called again. This torment continued until an exasperated Enrique picked up the hose, with water gushing forth over the patio floor, and managed to spray the parrot. That ended the parrot's shrieking—parrots are intelligent.

Tepic was a relatively quiet and sedate provincial town, or so it seemed to us, and we settled into a routine. We would drive down a steep street to the center to have lunch at the Hotel Imperial, where the waitresses were dressed in black with white aprons and looked old-fashioned, colonial, Castilian. Sometimes, after lunch, we had to wait until the street floods caused by midday rains had dispersed. There was one nightclub, and a story we were told revealed Tepic was not so quiet and sedate after all. Two men sitting opposite each other at a table got drunk one night. At one point one of them said, "I am better than you," to which the other replied, "Oh, no, you are not. I am better than you." The exchange escalated to the point that each simultaneously pulled out a gun, fired, and fell dead! The story was told to us by the hotel owner the next day.

In the third year, Rosemary stayed at UBC to mark exams, and Frank Tompa and I drove down to Tepic and stayed in the Motel Cora. Then Frank returned, and Rosemary flew down, and we met up at Mazatlán. Once back in Tepic, we decided to drive to Puerto Vallarta. A paved road became a gravel road, which became a burro track through forest. After leaving a small village called Las Varas, we began to climb a hill, and the car decided it had had enough and stopped. Some local men (Huicholes) saw us and helped to push the car to the top, from where we glided down to a small hut. A very hospitable family invited us to have a meal and stay for the night in a round thatched hut. When we woke up the next morning, they had all gone, before we had the chance to thank them one last time. We left some money instead. They were agriculturalists and had probably left before dawn to attend to their crops.

Mercifully, the car started, but our perils were not over. The map showed we could cross the Río Ameca at a shallow point, and this we did with great trepidation. Rosemary waded in front of the car and guided me as I drove through the shallowest part. It was good to be back in Puerto Vallarta, but shame upon us for being so ill prepared: all we had taken on this epic journey was some water and a half dozen grapefruits. We made one last and uneventful trip to the islands. On the return journey to Tepic, the car had difficulty in the same sandy soil right by Las Varas and, once again, stopped. The mechanic knew his business and replaced a burned-out condenser with a new one, and we completed our journey to Tepic without further mishap.

The next mishap occurred on a trip to Guadalajara to have the car serviced. Two-thirds of the way there, we ran into what looked like a dust storm, created by road-repair machinery. I took my foot off the accelerator and coasted, whereas I should have immediately braked, because we ran straight into the back of a pickup truck. I have to admit another shameful mistake: for the first and only time in our lives, we had taken off our safety belts, because there was no traffic. The front of the VW was buckled, but otherwise the car was functional. However, Rosemary had hit her chin on the raised metal flange by the heating vent on the dashboard, and it gashed her. We sat staring at the word INTERNATIONAL on the back of the truck, feeling miserable. Eventually, a burly policeman arrived on a large white horse, and we were escorted to Guadalajara and the police station. "This man can speak Spanish excellently," explained the policeman to his boss, sitting behind the desk. "But when I speak to him, he asks me to slow down, and as everyone knows, speaking slowly completely ruins the beauty of the Spanish language!"

All this rhetoric and sentiment poured forth while my wife sat there with a bleeding chin. The upshot was, we were given a warning by the police chief. When I said my wife needed the best medical attention, he phoned the university and persuaded the dean of medicine, a surgeon, to supply the necessary stitches. We were told he had to break off giving a lecture to do so. All's well that ends well. No, not quite. Rosemary developed hepatitis, possibly from a contaminated needle, although more likely from restaurant food. Many years later she was found to have antibodies to the highly infectious hepatitis A. My childhood hepatitis (chapter 2) was diagnosed as the same type, in the same way.

Our last year in Vancouver was 1964, the year of the Tokyo Olympic Games. I had been playing hockey in Vancouver regularly in the winter and was chosen for the Canadian national team. This put me in a dilemma, because to qualify as a Canadian I had to join the Army Reserve and attend activities once a week for six months. Joining would have created a conflict with thesis preparation; moreover, Rosemary was pregnant (a pregnancy destined to be unsuccessful). I declined but have since had regrets because it would have been a wonderful ending to my hockey playing. The Canadian players acquitted themselves well, especially against the British.

Robert MacArthur's *Geographical Ecology* (1972, 1) begins, "To do science is to search for repeated patterns. . . . The best person to do this is the naturalist." Once a pattern has been detected, it needs to be interpreted, so the next step considers alternative explanations and attempts to distinguish among them. In my thesis research, I discovered that several of the passerine bird species had evolved larger beaks and longer legs on the Tres Marías islands than on the mainland, a few strikingly so. The beak-size pattern had been noted before in other parts of the world but had not been investigated. I found the trends were not a simple consequence of larger overall size; in fact, the island birds were more often slightly smaller than their mainland counterparts. Nor could they be tied to a different climate on the islands; summer temperature profiles are almost the same on the islands and mainland. Instead, information on

FIGURE 7.6. Receiving a PhD at the University of British Columbia from President J. B. Macdonald, April 1964.

diets and feeding behavior pointed to ecological differences, as expected from the reasoning of David Lack and others about the scarcity of competitors. I concluded that some of the island bird populations, in the relative absence of competitor species and predators, reached high densities, competed for food, and diversified their diets and feeding locations. The conclusions rested on a careful comparison of island and mainland habitats, which was a novel feature of the research, and unlike other similar studies, the research encompassed many species in the community of birds instead of just one or two.

In writing the thesis I had a huge amount of help from a young faculty member almost my age, Geoff Scudder. Udvardy was away on leave again, and Geoff and Ian Cowan took me under their respective wings. Two oral exams were conducted, one for the department and one for the university, in which I made short presentations of the thesis from notes on a card and then answered questions. The two exams went well, and after the second one, Dean Walter Gage, a kindly man known generally as Uncle Walter, gave me wise advice: "Throw away your cards, and use minimal notes when you lecture. Don't be afraid to do so, you speak well, you can do it!" I have more than a memory of receiving a PhD degree; I have a photograph of the chancellor putting a hood on me at the degree ceremony (Fig. 7.6). Nice, except that it was back to front!

I had no job to go to, and Cowan said I could stay on as a lecturer, and so could Rosemary. It was very tempting, but I was awaiting a call from Evelyn Hutchinson to tell me whether he had enough money to offer me a postdoctoral fellowship at Yale. The call finally came through in July; I would receive $4,000 for a fellowship, and Rosemary would receive $4,500 as a teaching and research assistant. What a huge relief! We packed our bags and drove eastward in our VW to a new life.

8

Postdoctoral Interlude at Yale

A POSTDOCTORAL fellowship year or two is a time to broaden horizons, think differently, learn new techniques and ways of doing research, and explore before the next move into a salaried career. I used my fellowship year to publish my PhD thesis and related research, and then to plan future research and perform one small piece of research at the Connecticut Agricultural Experiment Station just north of New Haven. Thesis research had left me wondering how to test my main conclusion experimentally. Strictly, this is an impossible task, as the conditions under which competition occurred in the past cannot be replicated in the present. Nonetheless, competition can be investigated as a contemporary process. I designed an experiment to find out whether two species of sparrows that lived in separate habitats might restrict each other to those habitats by competitively interacting in a small, enclosed arena. It was a logical extension of my thesis research—not an end in itself but essentially a pilot study designed to test the feasibility of experimental studies of competition between species.

The results convinced me that birds were not suitable for this kind of experiment, but that ground-dwelling lizards or mice would be better, because they could be confined in the more realistic setting of a large outdoor enclosure. I made a trip to a herpetological meeting in Kansas City, Missouri, to talk to an expert, Don Tinkle, for advice. The best he could suggest was a couple of ground-living *Holbrookia* species of lizards in the Southwest. However, instead of remaining in the United States after my fellowship, I moved north to McGill University in Canada for my first academic position and switched to local mice and voles.

At Yale I was based in the Bird Department of the Peabody Museum, and we lived about a mile away, at 558 Whitney Avenue. There had been a falling out between the ecologists and molecular biologists and subsequent departures of several whole-organism biologists, but the atmosphere in the museum was stimulating and convivial. A social group formed around Philip Ashmole, a fellow Englishman, who was the acting curator of birds. Lunches in the bird room were memorable for the lively conversations with Charles Reed (mammals), Charles Remington (insects) and graduate students Tom Lovejoy and Peter Ames.

The intellectual level in biology at Yale was definitely one notch, or more, higher than at UBC, and I was ready for it. We had some outstanding seminars, including one on chromosomal puffing in *Drosophila* by Hans Beermann, a stunning tour de force from Donald Kennedy on crayfish neurobiology, and an equally impressive, quick-fire seminar by Richard Levins on ecological modeling of populations.

Dillon Ripley had left the curatorship of birds in the Peabody Museum to become secretary of the Smithsonian Institution. Three candidates for the vacant position gave interview lectures: Robert MacArthur, Robert Selander, and Charles Sibley. We young people wanted the position to go to MacArthur. I had had excellent interactions with him as we had similar interests in competition. He approached the subject as a theorist grounded in field observations, whereas my approach was more empirical and experimental, conceptual rather than mathematical, and the intersection of our views was highly stimulating. According to rumor, he would come to Yale if Dick Levins was also hired. Instead, Yale hired Sibley from Cornell University—probably a good choice for the museum but not a popular one, because he was inclined to be prickly and domineering. Our Cornell colleagues sent us a wry, waspish congratulatory message: "Yale's loss is Cornell's gain!"

Evelyn Hutchinson was writing the second volume of his monumental *Treatise on Limnology*, and so I was not able to interact with him as much as I would have liked. He was extraordinarily erudite across a wide spectrum of disciplines, and although his intellect was intimidating, his manner was gentle and his voice reassuring as he drew us into the exciting world of scholarship and ideas. I was greatly stimulated by his ability to make connections between disparate facts and ideas. From him I learned to be intellectually adventurous and bold, to have the confidence to stray beyond the safety of the immediately verifiable, and to be conceptually creative in exploring the world. He encouraged all of us to "think big," more by his example than by instruction. "Ask big questions and you *may* get big answers," someone (perhaps Evelyn himself) wrote, "but ask small questions and you *will* get only small answers." I can still see Evelyn's hooded eyes and sagacious gaze across the audience in the main lecture room as we waited for a seminar to begin.

Born into an academic family in Cambridge (UK)—his father was a mineralogist—and educated there in the classics and biology, Hutchinson was an emblem of the post-Edwardian era. He gave a stimulating graduate-level course on exobiology, in which we took it in turns to discuss the possibility of life on another planet. I suggested life might exist on one of Jupiter's moons, such as Europa, but he dismissed it as being too cold. However, with the discovery of salty water on Europa beneath a great thickness of ice, the idea is now being seriously considered by scientists. Once I had a long and interesting conversation with him when we sat next to each other at a dinner celebrating a birthday. At one point I gave a little discourse on the strangeness, from a modern point of view, of the Aztec calendar of twenty sets of thirteen days and later realized he had not added anything, which meant either that it was all new to him or, more likely, none of it was new, and he was too polite to say so.

I learned more about Hutchinson when I read his autobiography, *The Kindly Fruits of the Earth* (Hutchinson 1979). Intellectual interests encouraged by his parents contrasted with my sporting interests, also paternally encouraged. At cricket he "excelled negatively," while I did positively. Thus, our respective upbringings could not have been more different in the earliest stages, apart from shared interests in natural history, but they gradually converged, first through public school (his Gresham's, mine Whitgift) and then at Cambridge University.

Preparation for the adult world was very different in our two cases. He felt he had to be an adequate chemist in order to become a good field biologist, and thus equipped, he found his life's vocation in the chemical limnology of South African lakes. My route to a professional life was through animal behavior—bird-watching and butterfly collecting—and owed nothing to chemistry, at which I excelled negatively. For example, in my chemistry practical exam at Cambridge, I misjudged the heating of acetone in an open test tube over a Bunsen burner and set fire to myself. You would think once was enough, but I managed to repeat the mistake. I am able to write this only because a teaching assistant threw a blanket over me, twice.

Evelyn's description of life as an undergraduate at Cambridge is remarkably similar to what I would have written amost forty years on, despite a different set of dons and different relationships with them as a result of the insider advantage that I lacked. His uncle was Arthur Shipley, coauthor of a famous textbook on zoology (Shipley and MacBride 1901), who became vice-chancellor of the university. Evelyn describes him as "a rotund figure with one essential button not fully adjusted" (Hutchinson 1979, 62). In a world of zippers, who younger than fifty would understand the elliptical reference to trouser-fly buttons? It reminds me of a comical incident a few days before I emigrated to Canada. I was standing in front of a butcher's shop in Soho admiring the delicious meat when a policeman quietly approached, stood next to me, and while apparently looking at the same delicious meat, said, "Excuse me, sir, no offense, but your flies are undone." As I came out of my reverie, I said the first thought that came to mind, which was, "Oh, that's all right, they are always coming undone!" Not amused, the policeman edged closer, repeated himself, louder and with fewer words, and I mumbled an apology and moved to the shop entrance, where I discreetly attended to the button problem. Zippers must have swept through the market and replaced buttons shortly thereafter.

Our stay in New Haven ended with the birth of our first daughter, Nicola, in the Grace–New Haven Hospital. When Rosemary was two weeks past her due date, she telephoned me at the museum to say that contractions were coming at four-minute intervals. Strangely, I was not prepared for news that seemed to foretell a potentially disastrous unaccompanied birth. I ran the mile home in under four minutes, I am sure, to find Rosemary calm and smiling, already packed and ready to go, without having had any more warnings. I could not calm down instantly and managed to drive down a one-way street in the wrong direction to the hospital—fortunately, without incident. The receptionist's welcome only alarmed me more: "Please take a seat, and we will be with you momentarily." I then had plenty of time to calm down, as Nicola was born about five hours later. And what an amazing, indescribable feeling I experienced when holding this infant to my chest. It's not true that women have all the parental hormones.

9

McGill and Montreal

OUR SECOND DAUGHTER, Thalia, was born less than two years after we arrived in Montreal. We lived in an apartment on Ridgewood Avenue in the Côte-des-Neiges area, then a different one, and later transferred to first one and then another on Grosvenor Avenue in Westmount. The best was our last, number 466, in an old gray building three stories high; we occupied the ground-level apartment, and the landlord and landlady lived above us (Fig. 9.1). The children learned French with a French accent from French nuns at the Sainte-Marcelline (formerly Villa Marcelline) and then everything else at The Study. The latter was an excellent private school, especially good in English literature, and we felt our money was well spent. To give an idea of the spirit of self-confidence the school possessed, on a return journey several years later, we read a large banner outside the main building that proclaimed: *Le monde a besoin de femmes formidables* (The world needs great/strong women). Yes, it was an all-girls school. And Montreal was an excellent place to live with young children, except for one thing: a separatist movement (Front de libération du Québec, or FLQ) had become violent.

In 1970 a cell of the FLQ, dedicated to the independence of Quebec from the rest of Canada, captured Pierre Laporte, deputy premier of the province. Not long afterward, we learned they had murdered him. Another cell captured James Cross, a British diplomat and trade commissioner. He was rescued through negotiations many weeks later. It was a worrying time of uncertainty. Prime Minister Pierre Trudeau invoked the War Measures Act, which enabled him to station soldiers in

FIGURE 9.1. With Nicola and Thalia in Montreal, 1967.

Montreal, because the FLQ was threatening an insurrection. On the way home from school our children passed soldiers, standing on the street at strategic positions, and Nicola and her friends cheekily tried to get them to smile. Inevitably, there were grumblings from a small minority of people, who complained of overreaction on the part of Trudeau and unjustified loss of freedom. The perpetrators of the crimes were captured, exiled to Cuba, and many years later paroled and returned to Canada. Nonviolent activity was a more successful method of achieving most of the francophone political aspirations. A referendum was held in 1980 on the future of Quebec, and the separatist party (Parti Québécois), with Réné Lévesque at the helm, failed to achieve its goal of democratic endorsement of separation from the rest of Canada; a second referendum in 1995 failed by a smaller margin, and Quebec has remained in Canada ever since.

Two other large events dominated those early days. The first was Canada's celebration of its one hundredth birthday: Expo 67, the international exhibition, or World's Fair, centered on Saint Helen's Island (Île Sainte-Hélène), in the middle of the Saint Lawrence River in Montreal. The time was opportune to advertise the cosmopolitanism of Montreal, when English-speaking Quebecers were taking small steps to learn French and interact with their francophone counterparts. We visited the site several times, and I bought a balalaika at the Russian pavilion. We also took advantage of an artistic program associated with Expo that featured many celebrities. In one outstanding concert, Nathan Milstein played the Brahms violin concerto with an exhilarating virtuosity that reached its peak in a stunning, complex cadenza, impossible to describe. The best of several plays was *Uncle Vanya*, with Laurence Olivier.

Many friends came to stay with us during Expo 67, even from as far away as Australia, some having suddenly "rediscovered" us. My auntie Vi came for a week from England, as did my mother on another occasion. We measured the visits in bed-nights—120, I believe, but it seemed like 220. At the end of the summer, we escaped by car to Prince Edward Island for a once-and-only "ice-cream holiday."

The second event was a series of provocations from the student members of the McGill University Senate. This was in 1968, the time of Danny Cohn-Bendit, an anarchist leader of protest against the French government known as Danny the Red; the burning of tires in the streets of Paris; and talk of revolution. The protesting students at McGill were part of the ecology movement, the unintended sociological consequence of Rachel Carson's 1962 book *Silent Spring*, about pesticides, pollution, and environmental degradation. Our seven student senators learned by heart *Robert's Rules of Order*, a manual for conducting meetings, and ruined all meetings by claiming the right to speak, "on a point of order, Mr. Chairman." When one finished, another took over and thereby prevented any business from being done. It was a long time before this revolt in the misappropriated and misused name of ecology was brought under control.

McGill was a kind of Canadian equivalent to Yale, as a long-established, senior, urban university. The Department of Zoology was being rebuilt by David Steven, an import from Scotland. As at UBC, I felt I had elbow room to grow. Jaap Kalff and I were the young rebels, with galvanizing energy, applying critical rigor to everything. I was hired as an animal ecologist and taught the subject for the next twelve years. Increasingly we ecologists—at maximum there were seven of us at McGill—were becoming aware of environmental degradation through pollution from industrial developments, pesticides and insecticides, and outright habitat destruction. I found myself recommending politics and environmental law to those students who sought guidance on career options outside academia. There was a limited need for more research, especially into the effects of DDT and other chemical contaminants, but a massive need for the political will to tackle the major problems created in the name of society's "progress." A couple of decades later I joined a large number of senior scientists in signing a document circulated by the Union of Concerned Scientists entitled "World Scientists' Warning to Humanity." The introduction made plain a problem already perceived by Darwin, Alfred Russel Wallace, Henry Walter Bates, and other nineteenth-century naturalist-scientists in their travels:

> Human beings and the world are on a collision course. Human activities inflict harsh and often irreversible damage on the environment and on critical resources. If not checked, many of our current practices put at serious risk the future that we wish for human society and the plant and animal kingdoms, and may so alter the living world that it will be unable to sustain life in the manner that we know. Fundamental changes are urgent if we are to avoid the collision our present course will bring about.

Among the calls for action was this: "A new ethic is required—a new attitude towards discharging our responsibility for caring for ourselves and the earth." As I write, the global situation is far worse than it was thirty years ago. Some of the problems have arisen from the effects of anthropogenically caused climate change and the increasing frequency and intensity of extreme weather. Thirty years ago, the global average temperature was half a degree Celsius higher than the pre-industrial level. The increase in global temperature has now doubled, and the rate of increase shows no sign of slowing down. Coupled with climate change are multifarious problems that stem from increasing population density, associated resource needs, and poor regulation of how we use our environment. The world population has almost quadrupled since I was born—yes, a fourfold increase. But there is a glimmer of hope in that a new ethic to deal with the problem of caring for the planet is slowly, very slowly, developing. The critical question is whether it is growing fast enough—for humanity in general, and for Rosemary's and my children and grandchildren in particular.

McGill has a field station at Mont Saint-Hilaire, seventeen miles southeast of the city. It is a large, wooded property with a central lake (Lac Hertel), bequeathed to the university by Brigadier Andrew Hamilton Gault and known as the Gault Estate. I took students there for afternoon field exercises of sampling populations of mice, arthropods, and soil organisms, and we brought back samples for identification and conducting experiments in the lab. Later, in collaboration with my colleagues, I initiated a hugely popular two-week field course. The university had built a very comfortable building at the field station, with dormitories and kitchen facilities where the students stayed, but on most visits Rosemary, I, and the children stayed in a delightful old cabin-hut, notable for the head of a large elk stuck on the wall above the stone fireplace. We went there *en famille* as often as we could for weekends (B. R. Grant MS) and at maximum for one month when I was doing research on mice in the oak-maple-beech woodland. Gradually these amenities became known more widely in the university, more popular, more crowded, and more competed for. This is a theme that has been repeated several times in our lives.

The on-site director was Pat Baird, a tall, slim, upright, and soft-spoken retired military man who was a geographer. He and his wife were very hospitable; so too was Alice Johanssen, his replacement, although to our great regret she moved out of the main house and took over the cabin-hut. Before she did that, she introduced us one Christmastime to her father, the legendary "Jackrabbit" Johanssen. A pioneer, he had emigrated from Norway with his parents at the age of eighteen, worked in logging and geological camps until he was fifty, and then retired. He lived for another sixty-one years! He was famous in eastern Canada for introducing downhill skiing to improve the skills of his companion field-workers.

We met him when he was about ninety-five, and we could see why he had lived so long. For one thing, he was curious about everything, asking our children all sorts of questions about their outdoor experiences and enchanting them as well as us by showing how to make fire in the wilderness with a piece of lint and a flint. Another reason he lived so long was he was exceptionally fit. He introduced a New Year's Day skiing event from Montreal to Ottawa, a distance of about 120 miles, and invariably beat everyone else. At the age of ninety-seven, he fell down and broke a leg, and we all thought that was the end of his skiing, but we were wrong. He was back up on his legs the next year. At the age of one hundred, he started the race as usual and completed the first ten miles. The king of Norway invited him back to Norway that year and gave him a free tour of the country. He was gradually declining in health, and his eyesight was going, so Alice put him in a home for comfort and care. Even so, he was back in Norway at the age of 111, feted, and treated with great care. The end came when he said he was feeling a little tired and was taken to a local hospital, where he died soon after.

As the children grew, so did our horizons and our travels (Figs. 9.2 and 9.3). Following a brief holiday at Mývatn in Iceland in 1966, the first trip we made in Canada was to Prince Edward Island in 1967, and the next was to Newfoundland the following summer. The purpose of the Newfoundland trip was to combine research with pleasure, and it was so successful we continued with the same formula for the next decade in different places. The research goal was to collect voles for experiments on habitat choice back in the lab at McGill (next chapter). On the Newfoundland trip, we were helped by Rick Riewe, a graduate student from the University of Manitoba, and his wife, Jane, from their isolated camp on Inspector Island. The pleasure part was a trip north to visit a recently discovered (1960) and excavated eleventh-century Viking site at L'Anse aux Meadows, at the northern tip of Newfoundland. It is the only one of its kind in North America. By good fortune, Dr. Helge Ingstad, the Norwegian explorer who had discovered it (and together with his wife, archaeologist Anne Stine Ingstad, excavated it), had just arrived at the

FIGURE 9.2. The family at Thingvellir, Iceland, September 3, 1969.

start of a new field season. He graciously spent time explaining the lives of the Vikings as inferred from the archaeological finds—nails, a beautiful cloak pin, and a soapstone spindle piece—and easily imagined in the reconstructed Icelandic-style turf houses. We could not have asked for anything better.

Venturing farther afield, we twice drove to the Rocky Mountains in the early 1970s. I set up a small table and chair in a tent and spent the morning analyzing research data and writing. In afternoons, we took off on hikes. The weather was generally excellent, the skies a beautiful blue, and the air wonderful to breathe. We all loved the meadows, and I revived my boyhood passion for butterflies. The longest of the hikes was designed to see pikas (or rock rabbits, *Ochotona princeps*) and their hay piles on a rocky talus at high elevation on one of the mountains near the Columbia Icefield. We were camping in national parks midway between Jasper and Banff and well within reach of almost endless hikes into alpine meadows. In the second of the two years, we were ordered out of the Protection Mountain Campground after using up a two-week maximum allotment, which was a highly annoying bureaucratic reflex because there were not many campers at that time.

FIGURE 9.3. With Rosemary, Nicola, and Thalia, and Rosemary's father (**far left**), at Arnside, UK, 1971.

On one occasion a couple of old tourists approached our children and asked them a question. Rosemary watched at a distance and saw the tourists beat a hasty retreat, get into their car (with Florida plates), and drive off. Apparently, they had asked if there were any "bay-ers around hay-er," and Nicola, thinking they had said "birds" (rather than "bears"), said, "Oh, yes. It's easy to see them; they are everywhere!"

In 1974 I visited Australia to participate in the International Ornithological Congress in Canberra. It was a solo trip because we had no travel funds for the family. I joined a pre-congress tour to New Guinea that was guided by a very knowledgeable Australian, Lin Filewood. Papua New Guinea was about to become independent of Australian administration, and the main debating topic was its future name. Our visit was a bit chaotic; nevertheless, we saw more than two hundred species of birds, including bowerbirds and birds-of-paradise, and marsupial mammals. At the Baiyer River Sanctuary we were able to see captive birds and mammals close-up. While some members of our group were content to tick new species off their lists, my companion Tom Schoener and I were intrigued by how ecological roles were taken by completely different species from the ones we were familiar with in the Northern Hemisphere. An outstanding example is kingfishers, found there as forest-dwelling insectivorous birds living nowhere near streams. After the congress, many of us spent a few days at the Lamington Plateau in Queensland, again marveling at what to most of us were exotic birds in a fascinating piece of Gondwanaland forest. We stayed at O'Reilly's. Forty-two years later, I went back again, this time with Rosemary (chapter 24).

By 1975 we had spent ten years in Montreal. Before that we had never been in any one place for more than three years, so ten years was a long time, and we felt we had put our roots down. Despite the strong feeling of belonging, we had begun to feel a little restless. There were two reasons. First, Rosemary was frustrated, unable to advance her academic career by pursuing a PhD in the same department as me, because McGill had an anti-nepotism rule, and there were no obvious alternatives at other Montreal universities. Instead, she had done a teachers' training program in 1973 and had taught at a private school, Miss Edgar's and Miss Cramp's (B. R. Grant MS). It was a good school, notwithstanding the unexciting name, but once she got on top of the teaching, the future seemed to be one of routines and diminishing challenges. A second reason was a minor irritant. The department had hired Frank Rigler, a senior ecologist with a mildly hostile attitude toward evolution, which he considered a hindrance and not a help for a "true" (in his view, experimental) understanding of ecology. Furthermore, Gordon Maclachlan, a biochemist and the department chairman, once told me in all seriousness that evolution is not a subject to study, unlike genetics, because it had already happened and does not happen now. Attitudes like that discouraged if not foreclosed any growth in evolutionary biology in the department.

In the summer of 1976, we took off in our Volvo, destination Tucson, Arizona, where Robert M'Closkey, a friend on sabbatical leave, offered his apartment for a couple of weeks. On the way there we camped in Nebraska on the Fourth of July, and while the two hundredth anniversary of American independence was celebrated all through the night, we slept. Joking that we might be considered the old enemy, we kept a low profile until dawn when, with drunken bodies and signs of revelry lying scattered about the campground, we quietly left. Scenically, the North Rim of the Grand Canyon was the most spectacular part of the trip. The only blemish on an otherwise stimulating visit to this marvelous national treasure was Nicola's bad mood, directed specifically to me, which was funny because it was her eleventh birthday, and Arnold Gesell had warned me that from the eleventh birthday onward, a daughter might turn a bit hostile toward her father (Gesell et al. 1974). It happened right on cue, and then the clouds passed. As we continued south, we enjoyed the deserts of the Southwest, and not even rain all day at Meteor Crater, Arizona, could dampen our spirits. Everything was new to the girls, and we were excited because they were excited. From Tucson we made a camping trip into the Chiricahua Mountains, close to the Mexican border. A small rattlesnake sheltered under the fly sheet near the entrance, and we chose to think of it as more symbolic, as a guard, like Quetzalcoatl, than a threat to our health and welfare.

The summer trip nudged us toward applying for jobs in the United States. I had previously written to our very good friend Jamie Smith and asked him whether there was any chance of a professorship at UBC and was told, unfortunately, no. Had there been a position, we would have gone and would probably be there today. Our attitude toward the future was ambivalent, well illustrated by the fact that at the same time that I applied for a couple of positions I began the process of applying for Canadian citizenship. Rosemary was too busy with work to take the necessary steps, which she deeply regretted later, so it was I alone who was called for an interview in Montreal with a justice of peace, Hélène Baillargeon, otherwise well known for her popular bilingual TV program for children, *Chez Hélène*. I was most impressed by her wisdom and evident compassion. She explained that Canada welcomed diversity and encouraged new citizens not to give up their past culture but to contribute their experiences from elsewhere to society and so make Canada a richer community. I was both moved and proud to become a Canadian. This was in early 1977, almost seventeen years after I began graduate work at UBC.

During the application process, I received a phone call from Don Tinkle at the University of Michigan, the man who had advised me to study *Holbrookia* lizards many years before. He asked me to apply for a position at Michigan. With an attitude of "nothing to lose," I did so, was interviewed, and was offered the job after their first choice declined. This put us in a real dilemma for many reasons, not the least being the uncomfortable feeling that I would be betraying my recent Canadian citizenship by leaving. We wrote all the reasons for accepting the offer on the left side of a yellow notepad, and the reasons for staying in Canada on the right side. They were approximately evenly balanced. Then our cat, Gingy, walked across it—I should say marched across it—and left his paw marks as a signature on the left side, indicating his approval to leave. It took us a few more days to decide to do just that.

Much later, in the year 2000, Rosemary and I received honorary degrees from McGill (Fig. 9.4). It was a wonderful occasion, full of nostalgia and excitement, and we were thrilled by the experience, in my case of being invited back to my first university of employment, an

FIGURE 9.4. Receiving honorary degrees at McGill University, Montreal, June 2000.

emotional postscript to a vital chapter in our lives. I gained so much from those thirteen years of employment and the opportunity to grow as a research biologist and educator. For Rosemary, the award was different. She had enjoyed a more distanced attachment to the university, especially benefiting from the use of the library every Monday

(B. R. Grant MS). We each paid tribute to the university in our convocation addresses, and I ended mine with a nonstandard encouragement
to the graduating class: "You are, and will remain, for the rest of your
lives, the distinguished class of oo, and there will be no other o class in
your lifetimes. To a population biologist, the o class is the youngest.
Thus, by graduating this year, you have become blessed with eternal
youth! Make the most of it!"

10

Of Mice and Birds

A SIGNIFICANT fraction of our family life in Montreal had revolved around my research, as I will describe in this chapter. Here is the background. In my PhD thesis I had presented an argument with evidence for competition between species for food as a selective agent responsible for their evolution in the past. The next logical step in my thinking was to test experimentally the idea of competitive interactions in the present—to test a hypothesized process as an explanation for an observed pattern. At Yale I decided that competition for space was much easier to measure than competition for food and designed a study with mice. Joe Connell had done the only relevant experimental study, with barnacles (Connell 1961), and as a student wryly commented, "What kind of competition is it when one species grows over the other and smothers it to death?" Mice and voles are better than birds for experiments because they can be confined to enclosures. Generally, woodland species stay in woodland, and grassland species stay in grassland, but there are a few examples on islands where the one species present occupies the habitat of the missing species, regardless of whether it is the woodland or the grassland species. The inference from these observations is that they constrain each other's use of habitat when together, probably aggressively, and they are released from such constraints when they are on their own. I wanted to test the inference to see if it was correct.

I did this at the Morgan Arboretum of McGill University's Macdonald Campus, twenty miles west of Montreal at Sainte-Anne-de-Bellevue. The arboretum provided the woodland, and Dr. Louis Johnson, a

FIGURE 10.1. Enclosure for experiments with mice and voles, Morgan Arboretum, Macdonald Campus of McGill University, Sainte-Anne-de-Bellevue, Quebec, Canada, October 1966.

friendly farmer, provided an adjacent field. With the help of two students and a backhoe, we built three one-acre enclosures with metal fences, each of which comprised half woodland and half grassland (Fig. 10.1). We ran an electrified wire around the top to keep out destructive raccoons. We introduced only woodland voles (*Clethrionomys gapperi*) to one enclosure, only grassland voles (*Microtus pennsylvanicus*) to another, and both to a third, then let the experiment run by using live traps to find out where the animals went. As expected, they did not enter each other's habitat when it was occupied but did so in the absence of the other species after their own numbers had built up through plentiful breeding.

In a second and then a third year, I repeated the experiment using different enclosures for the two-species and one-species treatments and substituted woodland mice (*Peromyscus maniculatus*) for woodland voles because the latter were scarce in the source population at Mont Saint-Hilaire. The results were exactly the same as before.

I would have liked to perform the experiments with many smaller enclosures, to replicate them adequately for statistical purposes, but it would have been prohibitively expensive; moreover, a much-reduced area of woodland in each enclosure would have been insufficient for maintaining a population of woodland mice or voles because they normally occur at low density. Nonetheless, the experiments provided a successful test of the competition hypothesis, consistent with observations of mice and vole behavior in the laboratory (Grant 1972), and they encouraged me to turn back to the original question of whether competition is the explanation for some morphological trends in island birds, a subject that was becoming increasingly contentitious in the scientific literature (Schoener 1982).

The idea of linking evolutionary thinking more closely to ecology was in the air when I took a sixteen-month sabbatical leave in 1971–72, and I planned to write a book entitled "Evolutionary Ecology." But first I had research questions to pursue, and the book was never written, and a good thing too, because unknown to me, two other people had the same idea at the same time and published excellent books soon after (Emlen 1974; Pianka 1974).

I spent most of the sabbatical in the Animal Ecology Research Group in Oxford. We lived in a rented cottage in East Hanney, about eleven miles south of Oxford, close to Wantage and the North Downs. I had enjoyed the summer of 1969 as a visitor in Niko Tinbergen's Animal Behaviour Research Group in Oxford, and the stimulating conversations with Jamie Smith, John Krebs, and Mike Cullen. Together with the pleasures of renting a house in Kidlington, that experience had given us an appetite for much more. We relished being back in England and introducing the children, now four and six years old, to the many benefits that Oxford offers. Even though our rented brick cottage was damp, cold, and not well built, we had a garden for growing vegetables, making it so different from life in Montreal. The girls' grandparents enticed us frequently to the Lake District. I wrote papers and visited museums in Bonn, Vienna, Paris, Amsterdam, and Copenhagen to measure specimens of birds.

The sabbatical leave gave me the opportunity to study the classical case of character displacement. The two species of rock nuthatches

FIGURE 10.2. Model of an Eastern Rock Nuthatch (*Sitta tephronota*), near Shiraz, Iran, April 1972.

differ much more in the size of a black eye-stripe where they live in the same habitat in central Iran than where they occur separately, to the east and the west. The situation had been interpreted as the outcome of natural selection that minimized interbreeding. Differences in beak size could also be interpreted as a result of natural selection that reduced competition for food. The example of the nuthatches contributed to enthusiasm in the 1970s for the importance of interspecific competition in determining coexistence of closely related species. I was skeptical of the purported role of competition for food in this case and wanted to know whether the species differed in diet, and whether interactions in courtship had been involved in the divergence of the species in Iran.

In April I flew to Tehran and then to Shiraz. There I set out to see how each of the species responded in simulated behavioral encounters, not only with the other local species but also with its own conspecific relatives living alone, as in Greece or Afghanistan. To do this, I used museum specimens with or without their eye-stripes enhanced (Fig. 10.2), as well as playback of tape recordings of their songs. By and large the experiments showed that each species responded to the visual and acoustic cues of

its own species, and more to the geographically distant form that re-
sembled them than to the local form of the other species that did not.
This is the behavior expected if, in the past, the two species had come
together and diverged in the size of their eye-stripes. I also discovered
a strong difference in the seed component of their diets. However, this
did not correspond to a simple divergence in beak size; therefore, the
evidence for competition in the past was problematic. Together, the
results provided a better understanding of how character displacement
evolved. A modern study with molecular genetics is needed to probe
the evolutionary history of these fascinating birds to reconstruct the
steps by which divergence occurred.

In Shiraz, I stayed with an Englishman, Mark Taylor, and his Canadian wife, Linda. Each day I drove out into the country in a rented car
to experiment with the birds in a long canyon. On one occasion I returned to find some military people standing by my car. They indicated
I should get into my car and follow them to the nearest small town. Inside a police station, an English-speaking officer questioned me very
politely. He explained afterward that Iraqi terrorists had come across the
border and blown up bridges and military camps, which is why they had
been suspicious of me.

One day I took off as a tourist and drove to Persepolis, a most inspiring
huge step back into the past and unlike anything I had seen before. Built
on a raised platform in the fourth to sixth centuries BCE, the imposing
site stands conspicuously alone in a large plain and consists of very tall
limestone columns and the remains of massive stone buildings with
bas-reliefs of human figures, bulls and lions, and inscriptions. They are
like fossil fragments of an ancient world when the major powers in
conflict were the Greeks and the Persians—a disarticulated dinosaur
skeleton, *Tyrannus persicus*, with most parts missing. Persepolis was possibly used for ceremonial purposes until it was largely destroyed by
Alexander the Great.

I wish I had had the time to visit Isfahan as well, because in depictions
the mosque looks a marvel. As it was, I absorbed a lot of the pleasure of
living in Shiraz simply by walking around in afternoons. Before returning
to Oxford, I gave a seminar in the biology department of the university

and was taken with Mark to the Bamu Game Reserve, where we saw wild sheep, ibexes, asses, and antelopes, and signs of a recent kill by a Leopard. "Game reserve" meant a hunting preserve for the Shah.

In retrospect, I made a mistake in traveling to Iran without the family. Rosemary had wanted to get back into science by sharing the study of nuthatches and was thwarted on discovering that our plan to drive through Greece and Turkey to Iran and Afghanistan was not going to happen (B. R. Grant MS). To drive into Asiatic Turkey would have cost us more than two and a half times the price of our new Volvo, paid as a security bond to the insurance company, and that was well out of reach.

Instead, we drove to Istanbul and took advantage of a BP (British Petroleum) campground just outside the city gates to leave the car in a secure location and explore the city. Entering the city was like stepping into medieval times as we passed a bear chained to a stake at the city gate. Foremost among the sites were the Blue Mosque and Hagia Sophia. I had been introduced to them in an absorbing lecture by Nikolaus Pevsner in Cambridge, so it was a wonderful thrill to experience them in their architectural complexity and beauty, and to appreciate how they each differed from the Gothic cathedrals of Europe in attaining effects of heavenly spaciousness and lightness. We strolled through the wonderful bazaar of sacks of spices and dried fruits, a thousand things to look at and smell, and as we walked by the Bosporus and gazed across at Asia, our children were fascinated by the men smoking their hookah pipes. It was an adventure.

Before the Istanbul trip we had settled into a fairly new and well-laid-out camping ground at Kavala, in northern Greece. This was a huge piece of luck. While I continued the rock nuthatch experiments with the single species present in the low hills, the family explored, bathed in the Mediterranean, and did schoolwork. We arrived in early May and were there for more than a month. The aroma from the Mediterranean vegetation, the wild roses, the singing of linnets, all these stimulated the senses each morning as I climbed goat tracks to where the nuthatches were breeding.

Shopping for food was a challenge, since we knew little Greek, but unfazed, Rosemary responded to the challenge by imitating sheep (*baa*) and cows (*moo*) in the butcher's shop, precipitating a rapid exit of our two girls, embarrassed by their transmogrified but still earthbound ovine-bovine mother. Greece was under the control of the military, but you would not know it by the apparently irrepressible *joie de vivre* of the people; think of the 1964 film *Zorba the Greek* or the book on which it is based (Kazantzakis 1952). We bought fish freshly delivered to the wharf. The fishermen were overjoyed when they discovered that one of our daughters, Thalia, had a Greek name. "Hey, Thalia, Nicola," they shouted from a great distance along the wharf as we approached. What they thought of me I don't know. In Greek mythology, Thalia's mother was a mortal, and her father was Zeus.

Driving back through Yugoslavia, we camped outside the small town of Kotor, once Cattaro when Venetians ruled there. This too was a lovely experience for all of us. We met George, an itinerant and penniless Englishman as old as the century. He told us of his grim early life working in circuses. Food was scarce, he said, and many young children, forced to do daring acrobatics on tightropes high up under the canopy without a safety harness or net, fell to their death. George was, understandably, a socialist and communist. He told us tales of hiking alone in Albania, which was sealed off from almost all of us in the outside world. Since George had no food, and we had stocked up well for the long drive back through Yugoslavia, we fed him, and he repaid us with stories.

We were in Kotor for the final phase of the nuthatch research. I had the limited goal of tape-recording rock nuthatch songs to study how they vary from Yugoslavia to Iran and farther eastward. On the last day, before we drove out of the country, I kept hearing pinging noises above my head and realized after a while they always followed a *rat-tat-tat-tat* sound. Then I made a vital connnection: I was near a firing range, and bullets were ricocheting off the rocks! The sounds were on the tapes, and so was a conversation between police in Kavala inadvertently picked up from their radio transmission. Later we wondered what the border police would have made of them. So, the rock nuthatch work ended, as it began, with an encounter with the military.

On the outward journey to Greece, we had stopped in Slovenia and had a delightful walk in the rural countryside, at a time of year when a variety of butterflies had begun to appear in fields of flowers. We successfully resisted a strong temptation to make a side trip to sunny Dubrovnik and kept going. Apart from having hubcaps stolen for souvenirs by little boys in a small town in Serbia, we experienced no mishaps. No dallying on the way back, we drove the long journey through the lengths of Yugoslavia and northern Italy. We had used up most of our food and fistfuls of lire on the freeway to reach our destination—a campground close to the Swiss border—and when we got there, the English pound had just been devalued, and our travelers checks (in pounds) were unusable until the banks agreed on their value. The man and woman who ran the campground were our saviors: large, warm, compassionate, and generous, almost stereotypes from an opera. They could see we were in difficulties when we paid for our stay in coins and notes of five different currencies extricated from nooks and crannies in the car. The woman brought a heaping pile of ravioli that she said was "left over" from their supper, as a gift. Ravioli never tasted so good!

The next day we crossed into Switzerland, the pound was refloated, and so were we. We headed for a camp high up and away from the main road that looked interesting. The village was Zinal, and it was a superb choice, as the alpine flowers were magnificent in their vivid colors under a deep blue sky. Nicola celebrated her seventh birthday with a new red pack on her back. Even better, she somehow acquired a semi-tame chamois as a companion for the day.

This splendid interlude, part research and part holiday, ended when we returned to Montreal without a clear research program for the future but with a possibility. Ian Abbott had written from Australia to ask if I could fund him to do research on Darwin's finches in the Galápagos, his interest being, like mine, the role of ecological competition in evolution. I had thought of doing research on Darwin's finches ever since giving a seminar on them while I was a graduate student. More than a collection of bird species on islands, like the birds of the Tres Marías, Darwin's finches are related to each other through a common ancestor, which makes them ideal for studying the evolutionary origin and maintenance of diversity. But I had no money for studying them. However, Ian's letter pushed me into taking a risky decision, and I told him, yes, I could fund him on a fellowship, but we would have to search for money for the research. He accepted and came to Montreal with his wife, Lynette, a botanist.

We designed a program of studying six species of Darwin's ground finches for four months, aiming to gather ecological data to distinguish between contrasting hypotheses (with and without competition) to explain patterns of coexistence of species and differences in their beak- and body-size traits. Ian and I took opposite positions on competition for the purpose of dialectic debate. Ian was more skeptical than I was, as a result of having carefully analyzed the published data on Darwin's finches for a chapter in his PhD thesis, so our idea was to design the study in such a way that the data would convince each of us that the other was right. But where was the money to support it?

I asked the Chapman Memorial Fund of the American Museum of Natural History for half of our anticipated costs ($2,000), and our Faculty of Graduate Studies and Research for the other half. Eventually I learned the Chapman Fund would not support us because the research was fieldwork and not work in the museum. I told this to Walter Hitschfeld, dean of Graduate Studies and Research, over the phone and then added that we could achieve most of our goals with the half requested from the faculty. "You don't understand," he replied. "That is not the way it works. If we like your research, we will fund all of it. And if we don't like it, you will get nothing." And on that somber note, with hints of a Viennese background, down went his telephone. One week later, he called and told me the good news that the committee liked the research and would give us the whole $4,000. What a relief! I don't know what we would have done if the answer had been no. We had no plan B, for at that time I didn't know the National Geographic Society funded exploratory field trips.

The Abbotts started their field research in February 1973, just a few weeks before David Lack died, and also a few months after Robert MacArthur had died. We felt we had inherited something from both of them, a way of looking at nature and attempting to understand the patterns. Lack handed on a baton, and we ran with it. Armed with a newly published Galápagos flora (Wiggins and Porter 1971), the Abbotts spent two weeks at each of eight sites on a total of seven islands, capturing and measuring finches; banding them with a unique set of colored leg bands in order to recognize them after they had been released and when they were feeding; measuring their food supply; and systematically recording each item of food they consumed. An important item in their equipment was the McGill seedcracker. This was a handheld device for measuring the hardness of seeds that was designed and made for me by an engineering colleague at the university. I joined the field research for four weeks, two on Daphne Major Island and two at Bahía Borrero on the north shore of Santa Cruz Island. The experience was exhilarating from the first minute onward. Just as on the Tres Marías islands, the thrill came from the freedom to explore stimulating ideas in the complete isolation of wilderness, with nobody to disturb us, knowing that this small ecosystem was in an entirely natural state. All my senses were on high alert.

In no time, I was convinced it would be perfectly possible and safe for the family to accompany me on a follow-up visit. That took place in November and December. The purpose of our work was to compare finch communities measured at the end of the wet season (February–May) with the same communities toward the end of the dry season, in December, when food was not so plentiful. Rosemary took to the Galápagos as a duck takes to water, and the children did the same, like ducklings (Fig. 10.3; B. R. Grant MS). They were already experienced campers and were not fazed by a monotonous diet of porridge for breakfast, soup for lunch, and tuna and rice for dinner—or rice and tuna, for mock variety. It was our version of *My Family and Other Animals* (Durrell 1956), the story of a young boy growing up with his family on Corfu, Greece.

The result of making various comparisons was to convince us that the signature of interspecific competition could be seen in the composition of the finch communities, in the differences between species in their beak sizes and shapes and the differences in their diets that became magnified in the dry season when food was scarce, just as David Lack had argued and later Dolph Schluter demonstrated more comprehensively

FIGURE 10.3. **Upper:** With Jamie Smith (in hat), Rosemary, and Nicola (and a boatman) in a *panga*, Bahía Borrero, Santa Cruz Island, Galápagos, December 1973. **Lower:** Nicola and Thalia weighing a lizard at the Charles Darwin Research Station, December 1973.

(Grant 1986). But all was not accounted for by competition. The unique character of the food supply on each island was an important additional contributor to the patterns just by itself. We explained all this to Tjitte de Vries, acting director of the Charles Darwin Research Station in Puerto Ayora, on Santa Cruz Island. He listened without comment, and then said "Good. But you have to come again, because this is not a typical year!"

We knew there were particularly wet or unusually dry years, so already we were predisposed to return, and for another reason too. Our visits to Daphne had opened a door to an entirely new realm of research. At the beginning of research, we included Daphne because the Medium Ground Finch (*G. fortis*), in the absence of a breeding population of the Small Ground Finch (*G. fuliginosa*), was smaller there than anywhere else in the archipelago and had apparently taken over part of the ecological niche of the missing species by Lack's (1947) reasoning: a situation similar to the island voles described above. After the first year of fieldwork, I realized that little Daphne Major Island would be an excellent place to embark on a study of populations through time, and hence to study ecological processes that create the patterns we had documented and attempted to interpret.

Only eighty-four acres in circumscribed area, a volcanic tuff cone of four hundred feet elevation, with a central crater, clothed in a few trees and cactus bushes and in an entirely natural state, Daphne was then the home of two species of finches, the Medium Ground Finch (*Geospiza fortis*) and the Common Cactus Finch (*G. scandens*) (Fig. 10.4). On my first visit with the Abbotts in April, we had put color bands on the legs of many birds, here on Daphne and at other sites too. When we returned in November and visited five of the sites, we could scarcely find a banded bird at four of them. They had moved out of the local study areas or had died, whereas those on Daphne had nowhere to go unless they left the island, and we found more than 80 percent of the banded birds still alive.

This was a key discovery and a turning point in the research, for it encouraged us to believe we could study in time what anywhere else was only possible to study in space. It was thrilling, because the same juxtaposition of space and time gave Charles Darwin a key insight in

FIGURE 10.4. Three finch species on Daphne Island, Galápagos. **Upper**: *Geospiza fortis* (Medium Ground Finch) at a cactus flower (from Grant and Grant 2014, Fig. 2.19). **Middle**: *G. scandens* (Common Cactus Finch) cracking a cactus seed (from Grant and Grant 2014, Fig. 2.20; photo L. F. Keller). **Lower**: *G. magnirostris* (Large Ground Finch) cracking a *Tribulus* fruit (from Grant and Grant 2014, Fig. 6.10; photo K. T. Grant).

developing his theory of evolution by natural selection; in his case it was equating the separation of the ostrich-like Greater and Lesser Rheas in space (Patagonia) with a process of splitting from a shared ancestral species in time (Browne 1995; Keynes 2002): in a word, "speciation," the formation of two species from one.

"It is the fate of every voyager," wrote Darwin (1839, 474), reflecting ruefully on his departure from the Galápagos, "when he has just discovered what object in any place is more particularly worthy of his attention, to be hurried from it."

Fortunately, the modern naturalist has better options.

11

Evolution by Natural Selection

IN THE WORDS of Jacob Bronowski (1959), science is the search for unity in variety. Darwin gave us unity in the science of biological diversity with his theory of natural selection: "Here, then, I had at last got a theory by which to work" (autobiography in Barlow 1958, 120). Combined with Mendelian genetics and appropriate attention to the role of chance, the modern theory of natural (and sexual) selection gives us the tools to investigate and understand both the unity and the variety—that is, to understand the extraordinary biological richness of our planet. Darwin's finches are ideal for this purpose in several respects. They are a small group of closely related species that live in a varied and partly undisturbed environment, and none has become extinct through human activities.

The finches are an example of adaptive radiation—the relatively rapid multiplication of ecologically diverse species from an ancestral stock. The radiation is described as adaptive because different species exploit different food types efficiently by possessing beaks of different size and shape. They were made famous by Charles Darwin, who collected specimens that were described scientifically by John Gould, and then by Harold Swarth and later David Lack, who measured them and interpreted their evolution (Lack 1947). By using differences in their mitochondrial DNA, we now recognize eighteen species that diversified in the past million years from a common ancestor; seventeen on Galápagos and one on Cocos Island, Costa Rica, some 450 miles to the northeast (Lamichhaney et al. 2015).

Our research was initially motivated by a question of competition between species for food and then broadened to confront the more fundamental question of how two species are formed from one and coexist. The difficulty in understanding this process is not at the beginning, when two populations of the same species in different places diverge—chance factors are sufficient to cause divergence—but later, when they come together and coexist, with or without interbreeding. Thus, the key dynamic on each island is the interaction between closely related species, both directly and indirectly through their respective food supplies. Coupled with the speciation question is the problem of understanding why some populations are much more variable than others in beak morphology and body size. On Daphne, *Geospiza fortis*, a granivorous generalist, varies more than *G. scandens*, a cactus specialist. All these features made Darwin's finches an exciting group to work with.

Before moving to the University of Michigan in 1978, I took a sabbatical leave with my family in Europe and Galápagos. We spent the summer of 1977 in a rented house in Summertown, North Oxford, taking excursions to London and to Arnside to be with the Matchett family. Rosemary obtained from Oxford some very good teaching exercises to use with our daughters in Galápagos, and we bought books and visited museums in preparation for a long time away from such amenities and the benefits of society. Then we returned to Galápagos to begin a six-month period on uninhabited islands (Fig. 11.1) and to reap the many and wonderful rewards of combining family and professional life in truly splendid isolation (Fig. 11.2).

The first island was Daphne, where our research program was already underway and in its second year. *Geospiza fortis* and *scandens* differ in beak size and proportions and in body size. David Lack had interpreted such differences between pairs of species on several islands as the product of natural selection that reduced interspecific competition for food (ecology) and the likelihood of interbreeding (reproduction). Two married PhD students, Peter Boag and Laurene Ratcliffe, addressed the ecological and reproductive components, respectively. In 1976 Peter banded and measured many parents and their offspring and discovered from the strong family resemblances that all morphological traits of

FIGURE 11.1. With Rosemary and Thalia, looking at a map of Galápagos, Ann Arbor, Michigan, 1982. (Photo Jim Jagdfeld.)

FIGURE 11.2. Preparing for a trip to Daphne Island, Galápagos.

both species were highly heritable genetically and therefore capable of evolutionary change in contemporary time. He also found that the size and the hardness of seeds cracked by G. fortis individuals were strongly associated with the size of their beaks. Laurene found that each species responded strongly to beak- and body-size cues of its own species, and weakly to cues of another species. She did this by presenting a pair of museum specimens in lifelike courtship posture to occupants of finch territories. In a separate set of experiments with similar outcomes, she tested the finches' responses to playback of songs of their own and the other species, much as I had done in Iran and Greece, only more thoroughly. Thus, she learned that finches use both morphological and acoustic cues in discriminating between species and, presumably, when choosing a mate.

The next year (1977), there was a drought, and most of the birds died. This was a disaster for the second stage of Peter's heritability study, and of no help to Laurene either, but it turned out to be the most important event for many years. Natural selection occurred! Surviving G. fortis individuals had substantially larger beaks than those that died. The reason for the difference was this: only those with large beaks could crack open the large woody fruits of a plant called caltrop (Tribulus cistoides) to get the seeds inside, after most of the small and soft seeds on the ground had been consumed. Peter, in the middle of the drama, dutifully measured the diminishing food supply, all the while getting understandably distressed at the declining numbers of finches. Detached in Montreal, we were excited at the prospect of natural selection. When we returned to Daphne at the end of 1977, our major goal was to establish which birds had survived, and to find as many dead birds as possible. By now we were using numbered metal bands as well as colored bands for the study of parents and their offspring, so a dead banded bird discovered was an identified fatality. Nicola and Thalia proved to be especially adept at finding them—their eyes were younger and closer to the ground than ours—and helped us establish that mortality was not random but size-selective.

Because we had discovered in 1976 that beak-size variation in G. fortis was strongly heritable, natural selection in 1977 implied evolution in the next generation. We were compelled to continue, with the aid of student assistants every year since our own visits were restricted to summer months. Thus, a short-term investigation metamorphosed into a long-term research program. Years later we were able to demonstrate with numbers from the massive banding program that a detectable evolutionary change had indeed taken place as a result of natural selection. The fact that this happened in an entirely natural environment made it especially rewarding, because we could ignore any possible cryptic influence of humans.

Two PhD students helped to put these initial results in broader perspective. First, Trevor Price observed a similar but weaker selection event from 1981 to 1982 and found that selection sometimes differs among cohorts of the same species at the same time. This showed that selection in 1977 was not a once-in-a-blue-moon event but occurs repeatedly, albeit in varying strengths. Second, Lisle Gibbs discovered selection in the direction of small size when the food environment changed to a predominance of small seeds—thus in the opposite direction to selection in 1977. This is discussed more fully in chapter 14, and engagingly described by Jonathan Weiner in *The Beak of the Finch* (Weiner 1994).

All together, these findings were extraordinary because they showed that evolution could happen rapidly, repeatedly, and for reasons we could understand. The first important event that began in 1977 was an amazing piece of luck so early in our study and unlike any other study of natural populations at that time. Since then, numerous studies of vertebrate and invertebrate animal species have demonstrated natural selection, but few have been able to follow the evolutionary consequences of selection in the next generation. Moreover, the finch study was conducted in an entirely natural environment, with clear application to the broader picture of speciation and the generation of many species in the adaptive radiation of a variety of organisms, not just finches. In these respects, the study broke new ground and acted as a stimulus to the field. It opened up new possibilities that were previously thought to be out of reach.

We spent Christmas on Daphne under heavy leaden clouds, decorated a cactus bush colorfully, and pretended to have a Christmas dinner. This reminds me of an amusing story. Tourists used to come to the island and walk up a trail to a point where they could view the small crater below, standing with their backs to the place on the outer slope where we had our tents. In one such instance, the children were playing an imaginative game by the tents, and at one point Nicola said, "Oh, this tastes so good. It's just like roast booby." The tourists overheard this, took it seriously, and made a complaint to the research station and the national park office. On another occasion, a tourist complained that the metal bands were so heavy the birds could not lift their legs when they were flying, ignoring the fact that finches without bands on their legs do the same, dangling their legs and cooling them with the airflow. We survived these diplomatic incidents.

From Daphne we made a two-week trip to Pinta Island to help set up a graduate student, Dolph Schluter, and his assistants, Doug Nakashima and Eric Greene. Pinta was a new island for us, an attractive one too and more comfortable to camp on than Daphne. Introduced goats had not yet been removed, and they cropped the vegetation so much that it was relatively easy to make our way up to the top of the island and have lunch near a hot sulfurous steam vent. Camilo Calapucho, an old, seasoned field-worker, blazed the trail for us with a machete and placed rocks in the forks of trees as trail markers. Pinta has a different finch community from the ones we had experienced before, and the interlude there was fun and educational. From there we traveled eastward by fishing boat to Marchena Island, all day and very slowly, and then spent an hour late in the afternoon on the island recovering. The atmosphere was heavy, and the island was hot from internal warmth and enervating, unlike the energizing Pinta environment. On the following day, another long one, we completed the eastward journey to our destination, Darwin Bay on the beautiful island of Genovesa. This was to be home for the next three and a half months. Our time there was one of the all-time highlights of our family life.

Genovesa is a low, flat volcanic island, generally easy to walk on and covered with Palo Santo (*Bursera graveolens*), an aromatic torchwood tree. Viewed from above, the island is like a pancake, and Darwin Bay is a bite out of it. Nestled within the broad, curving sweep of the coastline

FIGURE 11.3. **Upper left**: Camp on Genovesa Island, Galápagos, 1991. **Lower left**: Tent on Genovesa Island, 1991. **Right**: Rosemary on Daphne Island, 2012. (*Le monde a besoin de femmes formidables*; see p. 85.)

is a small convenient beach for landing, and behind it is a small lagoon that fluctuates in area and depth with the tide. A wave-driven ridge of coral fragments separates beach from lagoon, and saltbush (*Cryptocarpus*) vegetation grows sparsely close to the ground. This is where we set up camp, behind the beach about one hundred feet or so from the shore, close to the lagoon and adjacent to the steep rocks that form the side of the bay; we rigged up an extra awning for shade and stored *chimbuzos* of water and food in various containers under a tarpaulin (Fig. 11.3).

We quickly settled into a routine. Until about ten in the morning, Rosemary and I worked with finches, and the children played or practiced on their violins (B. R. Grant MS). We returned to camp as it was becoming hot—this was the hot season—and under the shade we helped the children with their homework for about an hour, one-on-one. We split the subjects and taught the standard ones but had books on other subjects, including European art. Often we guided more than taught, and the children worked on their own, on mathematical problems, for example, or French translation (years later they could not remember this as teaching). Then we had lunch. After lunch Rosemary and I wrote up our field notes and occasionally read. Plays by Tom Stoppard were especially popular, thematically light and very witty, and we all read *The Lord of the Rings* and *The Silmarillion* by J. R. R. Tolkien. As the heat diminished, we left camp and either continued fieldwork or went for walks to the east beach, the west beach, or the center of the island to visit the partly saline Darwin Lake (Beebe 1924). Experience on Daphne had taught us to go to bed when it got dark and get up at first light, because we needed a long sleep after a thirteen-hour day. This was even more necessary in the heat of February, March, and April.

Some evenings we stayed up as late as seven p.m. Not quite ready for bed, Rosemary and I sat on a large balsa log riddled with holes made by the marine mollusk *Teredo* that had been pushed to the top of the beach by a very strong high tide. We stared at the gloriously bright stars and planets, or simply stared, and talked. On one such evening, night abruptly changed to day, and the cliffs across the bay glowed a bright golden color in what looked like late afternoon light. And just as swiftly night replaced day. We had to check with each other to make sure this really did happen. A meteoroid must have broken through the atmosphere and burned out in an incandescent farewell. On another memorable evening I started to fantasize about inventing a machine that determines the complete sequence of genes in the genome of a finch without even so much as touching the bird. I called the machine a genoscope, because it was like a telescope, and it worked by producing a laser-like beam that was reflected back to a receiver that would transmit the coded information via a satellite into the computer in my Princeton office. We are not there yet, but we do have complete genome sequences of all the Darwin's finch species from lab studies (chapter 26), and it's now up to the engineers to complete the conversion of fantasy into reality.

The lagoon was a focal point for relaxing. Working together, we turned a wooden pallet discarded from a commercial boat from who knows how far away into a *Kon-Tiki* and named it the *Sally Lightfoot*, inspired by the common shore crab of that name. The lagoon was also our swimming pool and a place for washing ourselves and our dishes. We shared it with the residents of the island, the Lava and Swallow-tailed Gulls, Yellow-crowned Night Herons and Lava Herons, sometimes Galápagos Pintail Ducks on an evening visit from Darwin Lake, and occasionally sea lions. Lava Herons were particularly interested in us. They learned that when we cleaned our porridge pot, fragments of food attracted small fish. Standing less than a yard away as we crouched on our haunches at the water's edge, and almost indifferent to our movement, they crouched on their haunches and waited until a fish came within range, and then struck. This was the special joy of living on Genovesa, being among the animals, almost in intimate contact, and being treated as if we were no more threatening to them than sea lions. I have had my toes nibbled by fish in the lagoon and pecked by a Lava Gull, and my right foot has been checked out by an octopus.

On one occasion on Marchena, I sat on the beach looking out to sea, and suddenly a hawk came from behind and perched on my head (Fig. 11.4). It even stayed there when I stood up! On another occasion, on Volcán Alcedo, Isabela Island, a large tortoise slowly and unconcernedly plodded past, within six feet of us, as we sat on the ground measuring and banding finches. Animals on islands unexposed to humans are well known to be tame. The pirates who frequented these islands experienced the same tameness three hundred years before us (Dampier 1927). It is one of numerous ways in which the Galápagos and their natural inhabitants are so special.

The seabirds had been studied during short-term visits to Genovesa (e.g., Nelson 1968), but apart from those, nobody had resided on the island for any length of time. A brief visit in 1973 had convinced me that it was possible to live safely with the family. We were lucky in so many ways to be there as a family in 1978. An occasional private yacht would call in on the way to the Marquesas, and when the travelers came ashore, they got a shock to find it inhabited by apparent castaways! Tourists visited rarely. Four boats were run by the main travel company, Metropolitan Touring, and they would come no more frequently than once a

FIGURE 11.4. **Upper**: A hawk on my head, Marchena Island, Galápagos, January 1988. **Lower**: Rosemary at banding site and tortoise, Volcán Alcedo, Isabela Island, Galápagos, 1999.

week and usually at longer intervals. When they came, there was no playing on the beach for us; the beach then belonged to the tourists. We often greeted them after they had finished sightseeing and were waiting for the Zodiacs to come to the beach and take them back to the boat. The guides welcomed a mini lecture to interested tourists on the research we were doing, and they helped us by bringing mail from the research station. Inevitably, the numbers of boats and tourists increased, and eventually our island lifestyle was no longer possible. Well before our research ended, so many boats wanted to visit the island that they had to book in advance, and the time spent ashore by the various tourist groups was reduced and strictly regulated.

The mood of the tourists was set by the spirit, enthusiasm, and responsible behavior of the guides, and with very few exceptions the guides set a high standard. Very, very rarely we encountered unhappy tourists. On one occasion Rosemary returned to camp to find a couple of doctors from Toronto just standing there and staring. They were clearly angry that we had brought the children to an uninhabited island, far from the nearest hospital. Rosemary explained we had carried one large bag entirely full of medical supplies with us, and she knew how to use them because she was a doctor's daughter. They left, scarcely mollified. In fact, we never had a serious fall or any illness. With water quality and diet fully under our control, we were as healthy as anywhere, and if someone had appendicitis or fell and broke a bone, we were equipped to deal with it. Admittedly, we had no phone or radio communication with the outside world. We reckoned we could survive an emergency until the next boat arrived.

The polar opposite to the disgruntled doctors was an eighty-year-old woman from Derbyshire, England, who used to work for the *Sunday Times* newspaper. Returning after work one day, we found her sitting in a chair at the top of the beach, facing inland. We were astonished to discover she had no legs! She had lost both of them many years before, and so she had to be carried ashore and placed in a chair at the top of the beach. She was in seventh heaven, she told us, the realization of a dream made possible financially by her family and physically by the extremely impressive, attentive, and strong guides. What a wonderful experience for her, and how heartwarming and inspiring it was for us to see someone realize a dream against all odds.

The research we did on Genovesa was like our work on Daphne, approached with the same sharp focus on the question of why populations of finches vary so much in their beaks and body size, and how the variation can be interpreted in terms of both genetic and environmental factors. The island is much more isolated than Daphne, being at the periphery of the archipelago, so we could be confident that birds were not likely to be flying in from somewhere else and affecting the variation by staying and breeding with the residents.

As on Daphne, we set about capturing, measuring, and banding birds, established who bred with whom, tape-recorded their songs, and banded the nestlings and captured and measured them when they had left the nest and had finished, or almost finished, growing. Focusing mainly on the Genovesa Cactus Finch (*Geospiza propinqua*; formerly *conirostris*), we compared offspring and parental measurements in order to estimate how much of the variation was genetically inherited. As on Daphne, we found the genetic component was very large, and we gained some insight into this by observing that the species exchanged genes (alleles) by hybridizing with the other two resident ground finch species, albeit rarely. This was valuable information, for it told us that any change in the genetically variable beaks caused by natural selection would produce a small evolutionary change in the next generation, again as on Daphne. Hybridization also told us that similar events on Daphne were not unique; they were just more pronounced and more easily studied on Daphne than on larger islands. To measure selection and the evolutionary response necessitated returning to Genovesa many times, and this we did for a decade.

As we pursued the study of finch families, we spent a large amount of time recording diets and estimating the relationship between the food available and the food consumed. In the breeding season, when caterpillars were abundant, all species fed on them, apparently unselectively, but later, when the supply of insects had diminished, the finch species switched to different feeding roles. It was very clear that their beaks—tools for gathering and dealing with food—defined their ecological niches or professions, and these in turn were key to their coexistence when food was scarce in the dry season. At this time their ecological niches were almost completely separate and nonoverlapping. Thus, one species, the Genovesa Sharp-beaked Ground Finch (*G. acutirostris*), was a generalist seedeater, constrained by its small size to eat small seeds. Another species, the Large Ground Finch (*G. magnirostris*), consumed large and hard-to-crack seeds such as those of Muyuyo (*Cordia lutea*). The intermediate third species, the large Genovesa Cactus Finch (*G. propinqua*), displayed its specialist habits of opening buds of prickly pear cactus (*Opuntia*) to feed on the pollen and probe the base for nectar, as well as hammering a hole in the side of a fruit to reach and extract the seeds inside—just like the related Common Cactus Finch (*G. scandens*) on Daphne. All this we recorded on banded birds of known identity and beak sizes.

Nicola and Thalia helped us in many of these activities while becoming interested in other birds as well. Nicola identified with the mockingbirds, fascinated by their family interactions and quarrels with their neighbors, and did a study of their group behavior. Sitting near their nest and observing the feeding of nestlings by banded adults, she discovered that sometimes three adults participated in the feedings: mother, father, and one other. The other, a helper, was usually a sibling from a previous nesting. This was a new discovery, and it provided the stimulus for PhD thesis research by my student Bob Curry. Thalia became attached to doves and made similar observations of nesting and feeding behavior. Later, with a little parental help, their written accounts became published papers.

We left Genovesa Island once to visit Wolf Island, which is situated remotely in the northwest of the archipelago. The splendid *Beagle II* from the Charles Darwin Research Station called in first at Daphne to collect Peter and Laurene, then at Genovesa for us, and took all of us to Pinta to pick up Dolph, Eric, and Doug, before continuing on the long journey to Wolf. We had plenty of experiences to share and compare, but Rosemary and Thalia did not take well to the boat's motion and lay down for the journey. We arrived on Thalia's eleventh birthday. She was in no mood for it but managed to rouse herself for a little celebration centered on a birthday cake made by Pancho the cook.

Wolf Island has steep cliffs and no beaches. Getting onto the island is an exercise in timing and good judgment. As the swell rises and falls, the necessary tactic is to wait until the sea and the boat reach the maximum height and then jump. There were plenty of hands and arms waiting for the children, and everything proceeded according to plan. All of us were in a state of relief, combined with exhilaration at being on such a remote island.

Most of the island is covered in a thick monoculture of small trees called Chala (*Croton scouleri*), but around the rim, boobies prevent their regeneration and thereby promote a growth of herbs and occasional cacti. This was the easy part for walking and working. We divided up the tasks and worked hard and collaboratively in our limited time, netting birds, finding nests, sampling seeds on the ground, watching and recording what the finches were doing, tape-recording their songs, and exploring. We kept our eyes open for the famous blood-drinking behavior of the Northern Sharp-beaked Ground Finch, sometimes called the Vampire Ground Finch (*G. septentrionalis*), although we were not there at the best time because it is more a dry-season than a wet-season habit. Nevertheless, Doug was lucky to see the bizarre habit once, a single finch pecking at the base of the tail feathers of a booby while the hapless booby incubated its eggs. At the end of the day, getting off the island was as demanding as getting on, perhaps even more so as the boat was rolling, and more than one adult staggered on impact. After we sailed away, Eric realized he had left his binoculars hanging on a branch. Almost miraculously, Pancho found them on a return visit a year later with another group of scientists. The binoculars were covered in guano. Eric sent a message to the research station, thanking Pancho and telling him to keep them as a souvenir.

A couple of weeks after our return to Genovesa, the time came to leave the wonderful island life we had enjoyed so much. The night before we were to depart, April 23, heavy rains fell for the first time in many weeks, and when the *San Juan* came to pick us up, we asked Bernardo

Gutiérrez, the captain, to go back empty-handed but return in one week. It was a valuable extra week, as we were able to see the effects of the downpour on the vegetation. The trees and shrubs had been losing leaves, and the breeding of finches was beginning to wind down. The rain reversed all that.

One week later we left, brown as berries, with the soles of our feet like leather from walking barefoot on hot coral fragments, and sad that our idyllic life had ended. Back at the research station we were greeted by friends, who said, "You must be glad to be back to have a shower and enjoy a beer." After twice telling the truth and being met with stares of incomprehension, I gave up and said, "Yes, it's great to be back." Well, in some ways it was, and in other ways it was not.

12

Michigan

WE ENTERED the United States at Detroit on the last day of August in 1978 and drove to Ann Arbor with our cat. Gingy had adopted us in the ferociously cold and snowy Montreal winter of 1971–72, and he was so much a part of the family he could not be turned back into the rat-catcher he once was. Besides, he was old. We set up home in a comfortable university house (300 Oakway) with a large fireplace, a feature that had been missing from our life in Montreal.

We had barely been in the house for a week when we narrowly escaped a disaster, one that could have been the worst of our lives. All faculty members of the Division of Biological Sciences were invited to an afternoon party at the house of Jim Cather, the chairman. We left Nicola and Thalia at home with Jim's phone number and instructions to go next door and ask to borrow the phone and call us, if necessary, never expecting the emergency that happened. Nicola was stung by a wasp. We rushed back home when she called, then on to the hospital, where she was immediately treated as an emergency. She was minutes away from a profound anaphylactic shock. A few days later we discovered from an allergist that Thalia was even more sensitive than Nicola. So began a long course of desensitizing treatment to take us out of the realm of nightmare.

Nicola went to an excellent private school, Greenhills, which had been set up a few years before by university parents dissatisfied with the education in the public schools and also with the permissive drug scene—Ann Arbor was in the avant-garde in its laissez-faire attitude toward drugs. Greenhills had the same advantages, of small classes and excellent teachers, as The Study in Montreal; moreover, it had the additional advantage of being coeducational. Nicola later told us she had culture shock twice a day, once when going to school and again when coming home. Thalia joined her one year later after a learn-nothing year in another school.

After one year we bought our first house, in a secluded dead-end street with access to a golf course through a small piece of woodland. A fireplace was a requirement. The garden contained a large patch for growing fruits and vegetables. It sloped down to a stream, unflatteringly known as the Pittsfield Drain, which, surprisingly, supported a population of crayfish. This was home, 1543 Stonehaven, and we could almost feel our roots go down.

Ann Arbor was a pleasant university town in which to raise a family, small, friendly, and convivial. On Saturday mornings we frequented the farmers market, socialized, and bought fresh produce. On one occasion we ran into Mrs. (Madame) Larson, the children's French teacher. She spoke only in French, as she did at school, yet directed most of the conversation not to the children but to us. How could she presume we understood, let alone spoke, French? I used to be able to speak French moderately fluently and was complimented once in Belgium on my excellent French accent, but I was rusty, and this ordeal came out of the blue. I struggled. Afterward, I received an unfavorable rating for my accent from our disdainful, and probably embarrassed, children. On the other hand, it was a warm experience of blending school with small-town social life.

We particularly enjoyed concerts in the Hill Auditorium and the Rackham Auditorium. The School of Music was one of the best in the country, and its well-recognized teachers were able to attract high-class professional performers and orchestras. The Vienna Philharmonic was a notable example. On one memorable Sunday afternoon, we listened to a piano recital by a budding star, Murray Perahia. I prefer listening to string instruments, but this performance was riveting, amazing for its power and expressiveness. For opera we went to Detroit.

I was hired as an ecologist and taught ecology. Gradually I took on behavioral and evolutionary aspects of ecology and taught them in graduate seminars. We had a rotating chairmanship system, and after a few years the time came for me to take the helm. Mike Martin was the chairman of the Division of Biological Sciences, and I was to be the chairman of the Department of Ecology and Evolutionary Biology within the division, which was staffed by about twenty faculty members. I asked Mike what it was like to be in the chair, and his memorable reply was, "Ninety-eight percent of it is ridiculously easy, and the other two percent is horrible." He was referring to time-wasting and disruptive problems caused by a small number of contrary faculty members. I experienced the same, although I had little to complain about because I was given very good support by most of my colleagues.

Our closest colleague was Bill Hamilton. Bill had been a contemporary at Cambridge, and although I saw him in zoology lab practicals, I don't think we ever exchanged more than an occasional pleasantry. I was apt to be chatty and gregarious in the lab; he was solitary and intense, a large head bent over a microscope. His theoretical work on kin selection and the evolution of social behavior made him a celebrity in Ann Arbor. He arrived at the same time as we did, and we had to solve similar problems in settling in. He had three daughters, Helen, Ruth, and Rowena (Rowie). They were a few years younger than our two children, who occasionally babysat for them. Our families bonded well.

Like us, Bill and his wife, Christine, hosted parties at their house, and guests were invited to stay overnight. One guest was the very distinguished geneticist Sewall Wright, then about ninety-three years old. He had given a talk in the Rackham lecture hall to an enormous audience. After one hour of addressing us on the history and development of population genetics, he had reached about 1924. Over the next forty minutes, refusing all polite and not quite so polite promptings to finish from the person who had introduced him, he reached 1932 and then stopped. Clearly, he was not lacking in stamina, despite his advanced age. Bill had a party for him in the evening, so we decided to arrive early, promptly at eight, because many people would want to talk to him. This proved to be wise. Standing in a small alcove and showing no sign of fatigue, he told us marvelous stories of spending the winter in Nebraska with a survey crew laying railroad tracks, and as he did so you could

FIGURE 12.1. In my office at the University of Michigan, Ann Arbor, September 1983.

imagine him as a robust, short and stocky young man. We talked a lot of science too, as more and more people came and joined us, most having to stand behind us. We left at about ten, with the chair we offered him still unoccupied. The next day Bill told us he continued like that until the last graduate student left at one a.m.

If Genovesa Island in 1978 was the high point of our life in Galápagos, my lab at the U of M was the high point of my professional life (Fig. 12.1). The interaction with postdoctoral fellows and graduate students was unbeatable. Dolph Schluter and Trevor Price had been accepted for graduate work by the department at McGill, but when I told them I was moving to Michigan, they both agreed to transfer there. They, Bob Curry, and Lisle Gibbs were the core members of the group, and I learned a lot from all of them. Dolph and Trevor gave us an identity, the Finch Investigation Unit (FIU). We used to have weekly discussions on anything of interest, a kind of journal club that I had started at McGill. Rosemary joined in, and we invited others on occasion to come for selected topics like group selection and evolutionary medicine. We also invited visiting seminar speakers to participate: George Williams, Earl Werner, Joe Felsenstein, and Bob Ricklefs, to mention just a few.

The department's strong reputation in ecology enabled us to attract outstanding graduates. Our discussions were lively, generally respectful, occasionally intense. Once while I was giving an informal talk in the Museum of Zoology, a graduate student could not restrain himself, and blurted out, "But that's not true, is it!" More embarrassing than that was my worst seminar ever. I was a prominent speaker at the first Midwest Population Biology Conference in Champaign-Urbana, Illinois. I was suffering from extremely painful neck and shoulders and could not concentrate on what I was saying. I fear part of it was gibberish, and I can hardly bear thinking about it now. Back in Ann Arbor, a doctor put a name to it, shingles, and prescribed simple relaxation as the best cure.

The year of the shingles, 1979, was stressful. We ran out of money at one point and had to get a bank loan of $3,000. We had never been in debt before, and this was especially worrisome for Rosemary. And I am sure I was working too hard. However, 1979 was also a momentous year that changed the direction of our lives, because a series of unexpected events brought our respective careers into even stronger alignment. It started with an invitation. A long and fruitful attachment to Sweden began when Staffan Ulfstrand at Uppsala University invited me to participate in a course for graduate students—actually, two one-week courses, back-to-back. The first course was on population biology of small mammals, co-taught with Michael Smith and David Pimentel, and the second was on bird evolution and ecology and was co-taught with Jared Diamond. The courses were held at a youth hostel outside Uppsala, at Erken. It was glorious springtime: the birch trees were clothed in fresh pale green leaves, migratory Pied Flycatchers were back, and Great Tits were singing their seesaw song.

The Erken experience was most enjoyable, in large part owing to Staffan, an effortlessly generous and courteous host. He shared my mischievous humor, and we got on so well that two years later I was back with the family for a whole semester. Staffan had been appointed professor at Uppsala in 1978, and in this position of influence he decided to make ecology in Sweden more international. He managed to persuade the Nordic Ecology Council to fund me for three months in Sweden and Norway. My mandate was to give a graduate course in Uppsala and then spend two weeks at Lund, Sweden, with Sam Erlinge, and another two in Oslo with Nils Christian Stenseth. Both were studying population dynamics of small mammals (voles and lemmings), a research interest similar to my own (chapter 10). In addition, my family and I managed to fit in brief visits to Copenhagen, Åbo (Turku), and Helsinki.

In time off in Finland, we visited Jean Sibelius's house in Turku and the largest fish market we had ever seen. Similarly, in Helsinki we visited the Rock Church, a strikingly beautiful church built directly out of solid rock partly below ground only ten years earlier. The low light of the sun's rays made the porphyritic crystals in the granite wall facings glow an extraordinary soft and warm purple color that matched the color of the carpets and upholstery. But I was suffering from a *Salmonella* infection that I had almost certainly contracted from delicious shrimp sandwiches on the ferry. How I got through the seminar the next day I don't know, nor how I was able to go out to the famous Troikka restaurant at the insistence of my hosts. They were adamant that I had to eat bear meat and drink an exquisite Armagnac for both pleasure and as a cure-all. Well, pleasure it certainly was, but the cure was much delayed. The train journey back to Turku was a black dream while rain poured down on one dark, sodden, plowed field after another. Yet, back on the ferry, I was miraculously better!

This visit to Scandinavia set the crucial stage for the next one. While we were in Uppsala, Staffan invited Rosemary to give a seminar on the Galápagos work and introduced her as "Dr. Rosemary Grant." Later in the evening, at dinner, Rosemary told him she was not a doctor, and a surprised Staffan replied that if she was interested in getting a PhD, he would be willing to be her supervisor (B. R. Grant MS). The answer was an emphatic *yes*. Rosemary had taken charge of the Genovesa research after the first two years, reoriented it, and published several papers, and had already explored the possibility of studying for a PhD at Michigan or an external PhD from Cambridge with Pat Bateson as supervisor, since her interests in sexual imprinting matched his. Neither came to anything, for reasons of money and time. The next day, Rosemary said, "Staffan, were you drunk last night when you offered to be my supervisor?" "No," he replied. "You have done most of the work already with your publications. All you would have to do is be in residence for a semester and pass an oral exam with me. It's free, but Peter will have to pay about two thousand dollars for a party for the whole department!" The deal was struck.

After Uppsala, we went to Lund. The weather in November was overcast, wet, cold, and blustery, in contrast to the sunny warmth of the people in the department. We stayed at the Stensoffa Field Station, about twelve miles from Lund, with the help of a loan of a Saab from a graduate student, and while I worked in the department, Rosemary and the girls enjoyed having the place to themselves and took daily walks in a woodland in the rain.

In Oslo we stayed at the main student residence. It was right next to a circular cross-country-skiing track illuminated by floodlights at night, and we took every opportunity to exercise ourselves. Nils and I had daily conversation about the biology of small mammals because at that time he was running a long-term study of a lemming population. Through his kindness we were allowed a week's holiday at Tömpte, once a country cottage of the king in the hills about forty-five miles south of Oslo and used by the biologists at the university as a field station. We were happily snowed in. For provisions we skied down the winding lane to a food shop on the corner of the main road, and then hauled the food back on a sledge. For warmth I got up before everyone and fed firewood into a stove that gave us radiating heat. Day length was not much more than four hours. Every day as the sun rose, a group of Black Grouse came up from the valley to feed on catkins in the tops of the birches, keeping just above the shadow line, and then in the early afternoon when the shadows lengthened, they flew back downhill. We left after a week, thankfully missing the annual pig slaughter at the adjacent farm.

On returning to Britain for Christmas, we rented a car and drove to Oxford to stay with Tom and Christine Getty, friends from Michigan. Disaster awaited us. We took some of our belongings into the house, locked the car, and went inside. When I went back to get the remaining bags half an hour later, I found a policeman standing by the front window of the car. It was broken. He explained that as we drove into Oxford, we were probably followed by a couple of thieves in a pickup truck, and when we all went into the house they drove around the block, returned to the car, and within one minute had literally carried out a smash-and-grab. One of my bags was saved by being wedged between the seats. The children had taken the violins into the house, but they lost all their clothes and school notes and, worst of all, their diaries. Rosemary lost all her best clothes and jewelry: a ruby and diamond engagement ring and an expensive and exquisite garnet and pearl brooch.

Next day we went shopping for clothes. The children shopped on their own, going from one shop to another, and when we all joined up, we found they had bought identical clothes. Christmas at Orchard Close was a most enjoyable way to forget we had ended our rewarding Scandinavian holiday on a bitter note of personal loss.

Back in Ann Arbor, it was a sad day when our cat died. He had appeared to be in excellent health up to the end, and even though old, he was canny enough to ambush chipmunks. Not knowing his origin, we guessed his age was about nineteen. He was an aristocrat, an Abyssinian with a recessive gene responsible for long hair, known as a Somali cat. He had belonged to the sister of Moshe Safdie, the architect who designed the revolutionary Habitat complex for Expo 67 in Montreal. After a while, she had lost interest in her cat and turfed him out in winter, where he survived on scraps of leftover vegetables thrown onto the snow by our neighbor and on rats. He chose us in preference to the rats. I was the last member of the family to hang on to our freedom from pets—for reasons of convenience, not because I disliked them. One morning when I was working at the living room table, the cat squeezed in beneath a barely open window, walked softly behind me into the bedroom, and fell fast asleep—on my pillow. That did it. For many years Gingy comforted our children with his bodily warmth, and many were the tears shed into his lovely ginger-colored fur. From his point of view, his best gifts to them were the mice he carried into their beds, some alive, some dead.

13

From Michigan to Princeton

HISTORY REPEATS ITSELF. We enjoyed Ann Arbor, never sought a position elsewhere, and as in Montreal, we thought we would retire there, but out of the blue came a telephone call from Bob May asking if I would be interested in the possibility of a move to Princeton, New Jersey. I would be the third prospective candidate. As a member of the Visiting Committee for the Biology Department at Princeton in 1982 and 1984, I had got to know the population biologists quite well—Bob, John Terborgh, Henry Horn and Dan Rubenstein—and could see them as intellectually lively, mutually respectful, and compatible colleagues. I had also close to an insider's view of the schisms between organismal and population biology on the one hand, and molecular biology on the other, amusingly described by John Bonner (2002). It was the same old story I had experienced at Yale, revisited. Ed Wilson experienced the same at Harvard (Wilson 1994). On the positive side, the Princeton University administration responded to the committee's strong recommendation to create a single department from the separate Biology and Molecular Biology Departments in order to reduce antagonisms. The change was a stepping-stone to further reorganization, but at the time it looked as if this was the right move to promote integrated biology at Princeton.

I gave an interview seminar and shortly after was offered the position. For Rosemary, there was a place on the research staff, her first university position. It was a major inducement. We returned as a family in springtime, when Princeton looked very attractive and welcoming, made more so by the warmth of our friends. The university was more like my Cambridge origin than anything else I had experienced in North America. The decision to accept the offer was far easier than the earlier decision to move to the United States from Canada. In neither case did the university I was leaving try to persuade me to stay, which simplified the decision. My colleague Dick Alexander's friendly comment was, "Princeton will suit you; you are the Ivy League type" (he was not). Before we took our decision, Evelyn Hutchinson came to the U of M to give a seminar, opening with the memorable words, "The older I get, the more interested I have become in the history of ecology, probably because I am part of it." He stayed with us at home, and I told him of the Princeton offer. What did he think of the population biologists, I asked, to which he gave a slightly sour endorsement: "They are good, they are very good, but not quite as good as they think they are." Ooh—perhaps there was a Yale-versus-Princeton ghost peeping out of the closet.

In any event, I accepted, and we went. But first I claimed a year's leave of absence, equivalent to the sabbatical I missed by leaving Michigan. My first sabbatical was so enjoyable, I joked on returning that I saw my career as a string of sabbaticals with lectures given to undergraduates in between. So I was not willing to give up this one. Selling our house in Ann Arbor was notable because houses were not selling. "You can't lose money on a house," was the prevailing wisdom in the year we bought it, but we discovered it was possible. We also sold our fourteen-year-old Volvo station wagon, with great reluctance, in order to buy a new one.

The vehicle was waiting for us at the factory in Gothenburg. Our good friends, fellow biologists Ulla and Åke Norberg, looked after us for a few days, and then we drove east to Uppsala to begin Rosemary's required semester in residence for her PhD degree. Previously we had lived near Gamla Uppsala and did a lot of walking into town and back in fading light when jackdaws were streaming in to roost in the cathedral. This time we lived at the top of Villa Åsen, the university's eighteenth-century guesthouse, close to both the department and the botanical garden. And unlike previous visits, we had our own car, the third Volvo station wagon we have owned, currently in its thirty-eighth year! It was part holiday and part work. The children were off at college,

Nicola at Dartmouth and Thalia at the University of California at Santa Cruz.

The climax of the semester was the buildup to Rosemary's thesis defense, the oral exam, and the "nailing of a thesis" (literally) to the university building—a very long tradition—to allow people to read it, object to it, and even question its authorship. Nobody objected, and Rosemary's thesis defense took place in November. Staffan had told Rosemary the defense had to be top quality because the audience was expecting a high level of critical, informative yet entertaining debate, and separately he told the external examiner, André Dhondt, the same. Thus primed, both candidate and opponent rose to the occasion; it was indeed a spirited debate.

The second climax of the day was the party in the evening. Our friends Ola Jennersten and Gunilla Rosenqvist guided us in preparing for it, because they knew what was expected. This included a bizarre drive the night before to the outskirts of Stockholm to collect a large ham from the freezer in Ola's brother's workplace. We were told we had to have it to accompany the reindeer meat, and we happily followed friendly instructions. All students and postdocs knew Rosemary well, and they made sure the party was a prolonged, lively, at times hilarious, and unforgettable event.

We spent the second half of this quasi-sabbatical at Oxford. I had previously enjoyed being in Niko Tinbergen's Animal Behaviour Research Group (1969) and in John Phillipson's Animal Ecology Research Group (1971–72). This time I was given space in the Edward Grey Institute of Field Ornithology by Chris Perrins. I occupied David Lack's office in a new building on South Parks Road, and while he was enjoying the heat in Jamaica on sabbatical leave, I nearly froze each day, and only an electric heater aimed at my feet prevented ice forming. I spent most of my time preparing a new set of lectures for the first semester at Princeton.

Thanks to Dick Southwood, chairman of the Department of Zoology, I had a visiting professorship in Jesus College, and we lived in a college flat on Woodstock Road. Like any other heterogeneous academic group, the fellows spanned the range from friendly normal men

and women to the strange and the weird. I was able to observe something of college life by having lunch there on most days, benefiting from
conversations with diverse scholars and gaining some insight into what
it must be like to live the life of an Oxford college fellow (a don). One
has to be an insider to know what it is truly like (e.g., Murray 2019). Up
to about the age of forty or so, I would have jumped at the chance to
become one myself, but by now, at forty-nine, I had adjusted to a very
different North American society and lifestyle and was not so sure that
it was such an attractive life. I did not feel alienated in the college, far
from it, and in most ways, I was still English and the product of my early
upbringing. I could act the part of a visiting fellow without acting; however, college life did seem somewhat artificial through eyes conditioned
by North American experience. I can understand Thomas Henry Huxley,
when he wrote to his son about declining a university position: "I do not
think I am cut out for the life of a Don nor your mother for a Doness—
we have had thirty years' freedom in London, and are too old to be put
in harness" (Huxley 1900).

Several years later, I was asked to allow my name to be entered as a
candidate for the vice-chancellorship at the Universities of York and
Southampton. I declined both. I had a similar invitation, this one by
phone, to be considered as the head of the Zoology Department of the
Natural History Museum in London. Bob May, a contemporary, had
recommended me. This was a civil service position, and I knew that the
retirement age was sixty. I had to tell an embarrassed man at the other
end of the phone that I had already reached that age. The only position
that might have been tempting was a professorship at Cambridge, because
it would have been an opportunity to return some of the benefits I had
received from an excellent education at both school and university. It
was nice of Pat Bateson, my friend from undergraduate days, to ask me
whether I "would be prepared to consider it," but the idea died when I
learned there would no position for Rosemary.

Bill Hamilton had left Ann Arbor the year before us to take up a
Royal Society Fellowship at Oxford. We saw quite a bit of him, Christine, and the girls at their house in Wytham. He did his best to help me
see my first European Badger late one evening by taking us to just the

right place at the right time, "where you almost always see them." I still have not seen one.

Bill also extended our experience by taking us as guests to his college, University. There I learned that Jesus College did not have all the socially strange fellows, in fact, relatively few. However, Rosemary had the good fortune to sit next to a very interesting man, and they had a lively conversation throughout the meal until Rosemary unthinkingly passed the port the wrong way. He was mortified and cut her dead. She was furious with him but more with herself, as her mother had taught her correct etiquette decades before. The next gaffe was mine. Dick Southwood took us to dinner at his college, Magdalene. We enjoyed everything except the conversation of one of the fellows, nicknamed Batty Barton. I reciprocated by inviting Dick and his wife to Jesus, and then I put my foot in it by failing to observe the correct protocol. Dick was vice-chancellor, and I introduced him to Peter North, master of the college, instead of introducing Peter to Dick according to their respective ranks in the hierarchy. I was forgiven, I think, by being semi-foreign and out of touch—in my native country!

On this and other visits to England, Orchard Close was the domestic focal point, more than London or Winchester, where my mother lived. We drove up to Arnside as often as we could. Early summer, before the roads become clogged with traffic, is ideal for long walks and exploration of the fells in the Lake District. We returned to favorite haunts—Tarn Hows, Wasdale, Coniston Old Man, and Sty Head Pass—and we became thoroughly English again. Before that, in April, we took a holiday in Nepal, where Nicola was spending a year abroad in Kathmandu in a program sponsored by the University of Wisconsin. She joined us for a few days in the Chitwan National Park and then, when she could not be with us any longer, we went trekking in the Himalayas.

Chitwan is a relatively open sal forest populated by tigers They stayed away from us, however we were able to see rhinos among the well-named elephant grass while seated on the backs of elephants, as well as jungle fowl, peacocks and peahens, and a bear. Entertainment one evening took the form of an interesting dance in which young men with a short stick in one hand danced vigorously to rhythmic music while slowly moving in a circle and watched by young women and older people. The exercise consisted of striking the stick of the man in front, then turning to strike the stick of the one behind, reversing, and so on, all the while gradually moving counterclockwise. This looked like a courtship dance in which females were able to compare the skills of the males and make a choice; or perhaps their parents did. Chitwan is an area of lowland Nepal where malaria is present, and so is a genetically based resistance in humans that differs from the better-known sickle-cell anemia trait. The dance is a test of stamina and would surely reveal the unfittest. We happily accepted the invitation to join in and experience the excitement of the rhythmic dance.

In Kathmandu we followed a recommendation to ask a trekking company if we could hire a guide called Pasang. Pasang is a common name, and by good fortune we were given Pasang Sherpa, a highly experienced Everest guide. He was about fifty years old, short, and wiry. Our goal was the Langtang Valley.

A bus drove us to Dhunche, where we took a break and strolled through the bazaar. Then we hiked according to a daily routine, typically for a few hours before a late-morning breakfast, followed by several more hours before stopping for the night at hotels or teahouses: the Yeti Hotel at Syabru; Kyang Jing, the site of a Swiss dairy; the Lama Hotel beyond Gora Tabela; and Sing Gompa. At one of our stops a merry teahouse owner laughed as he used his abacus to calculate our bill far faster than Pasang could on his calculator. We had just eaten a delicious and much-needed breakfast omelet.

A Swiss couple, Stefan Ziegler and Regula Bürgi, joined us, and for most of the trek we proceeded as a quintet. Pasang explained that he had helped a French anthropologist in this region to work out the pattern of marriages. A man had to walk for ten days before qualifying for marriage, which is an effective way to minimize marrying relatives. Along the way we passed through coniferous forest that was superficially like a North American or European forest except for an understory of rhododendrons and the bushes and small trees of *Daphne*, with lovely white flowers—yes, Daphne again. Sunlight at a low angle intensified the

FIGURE 13.1. At the head of the Langtang Valley, Nepal, with Rosemary and Pasang Sherpa, 1986.

bright and variegated colors of the rhododendron flowers, matched by colorful minivets as they flew through the treetops.

The head of the valley is a striking terminal moraine (Fig. 13.1). Temporarily and happily freed from our heavy packs when we arrived, we made the mistake of walking too quickly and succumbed briefly to the altitude when climbing it. On the final leg of the journey, we climbed out of the valley and hiked across open country in snow and ice above tree line to the frozen lake of Gosainkunda for a final overnight stop in a stone hut at about fifteen thousand feet. Pasang promised we would see a Red Panda on the way, and we did, briefly but clearly. It looked like a large member of the weasel family as it bounded across the trail in front of us to escape a harrassing crow and then disappeared. That night was cold, despite the fire, and we had to lend Pasang our reflective space blanket. A couple of small birds (Alpine Accentors) came into the hut for warmth.

Our return to Uppsala for convocation and the awarding of degrees took place in the summer of 1986 (Fig. 13.2). This was a lovely occasion for both of us; Rosemary received her earned degree, and I received an

FIGURE 13.2. Celebrating Rosemary's PhD and my honorary degree at Uppsala University, Sweden, with Staffan Ulfstrand, 1986.

honorary degree. She rented an academic gown with the dark green trim of her Edinburgh undergraduate one. Rosemary's admiring parents were in the audience, watching their daughter's crowning educational achievement. We each wear a Swedish ring as a perpetual memento, married academically, as it were, by Uppsala University.

I was hired as an ecologist for all three of my university positions. Princeton wanted someone working at the interface of population and community ecology, and that fitted me well. I am also an evolutionary biologist and was able to strengthen that dimension in a department that is distinguished in ecology primarily through the pathbreaking work of Robert MacArthur and Robert May. Henry Horn wrote to tell me that he and John Terborgh had been teaching evolution, and I should take over the course because the subject was more mine than theirs. That was also fine by me because I had been teaching evolutionary ecology in Ann Arbor. All this did was to give me a new label. At Princeton, I shared the course with Marty Kreitman, a new assistant professor who taught the population genetics and molecular genetics part. When he left a few years later, I added his part to mine. I never taught another course in ecology, as we had other ecologists to do that, including Henry, and two new faculty members, Simon Levin and Steve Pacala.

We occupied a university-owned house on College Road West. Like our house in Ann Arbor, it was next to a golf course and had a fireplace. The neighbors in our row of four houses were interestingly diverse: a mathematician, an Islamicist, and a classicist. Rosemary and I were the only ones to appreciate the snapping turtle in a small pond behind the house. The house was comfortable and a convenient short walk from the central part of campus and from Eno Hall, where Rosemary and I had adjacent offices. Although we knew we would want to buy a house, we were in no hurry and could see ourselves staying for as long as the university allowed us. We looked at some houses on the market and were not impressed.

Our first trip to Galápagos from Princeton occurred in 1986, a dry year of little rain, and the finches showed little interest in breeding. Prince Philip and his entourage from the United Kingdom paid a visit to Genovesa while were there. Responding to a prompt from Gunther Reck, the Charles Darwin Research Station director, I offered Prince Philip a copy of my brand-new book on Darwin's finches (Grant 1986). He thanked me and handed it back to me to autograph. As I stood there on the beach, dressed only in shorts, wondering how, as a loyal British subject, I should refer to him in the inscription without giving offense— HRH Prince Philip? Duke of Edinburgh? or simply Prince Philip?—he quickly ran out of patience and said: "What's the matter, have you forgotten my name?" And before I could explain, he added, "Perhaps you have forgotten your own name," and smilingly turned and walked away. I managed to inscribe and deliver the book before he left.

The trouble with casual encounters with royalty on tropical beaches of remote islands is there is no opportunity for rehearsal. Previously we had had an unexpected visit from Prince Bernhard of the Netherlands, surrounded by a half dozen muscular guards. Rosemary, dressed in boots and bikini, with little warning, was introduced to the poised and courteous prince. Her instant dilemma was: Does she bow—in a bikini? Or does she curtsy? He solved the problem for her by holding out his hand for a handshake.

The following year was the last full field season on Genovesa before we drew the study to a close and converted the annual, long-term research into a book (Grant and Grant 1989). We had decided to concentrate our efforts on Daphne because it was so much easier to find all the banded birds there. Our study area on Genovesa encompassed only 10 percent of the island. No ordinary year, 1987 was the second El Niño we experienced. The first was in 1983, when a record amount of rain fell (chapter 14). Somewhat less rain fell in 1987, but it was hotter, humid, and physically draining. By ten in the morning, putting one foot in front of the other was an effort. Nonetheless, it was enjoyable to be watching birds breeding for the first time in three years, and valuable in what our observations told us about survival and the breeding structure of the finch populations (Fig. 13.3). After that year, we returned to Genovesa Island a couple of times to monitor the populations and to get blood samples for DNA analysis (chapter 14). Walking for long stretches and seeing only birds without bands was a strange experience, rather like stepping right back to our arrival in 1978. On one of these visits, in 1994, we found a *G. propinqua* male still alive at a minimum age of thirteen years. This was 3044, a favorite and almost unique in singing both a fast and a slow version of the standard song instead of just one of them.

On returning to Princeton, we resumed the search for a house, without success. We were on the point of giving up, disheartened, when Judy McCaughan, our excellent realtor, encouraged us to take a look at a house that was beyond our price range. "One last house," she said. Normally, when visiting houses for sale, Rosemary and I walked around together and made comments, so on this occasion our mutual silence was loud and significant. We loved it, and not just because it had a fireplace. Sharing opinions as we walked for twenty minutes back to Eno Hall, we decided, at the end, to put in a bid. We were lucky to have it accepted, because two other prospective buyers had bid already, but we had the

FIGURE 13.3. Writing notes at Genovesa Island camp, above the beach, 1987.

advantage of a university mortgage; moreover, we had no obligation to sell a house first before paying for this one, unlike our competitors. The house on Riverside Drive became ours in July, thanks to a generous financing scheme from the university (Fig. 13.4). It is still home.

El Niño and a house purchase. As if these were not enough, there was one additional outstanding experience awaiting us: the emergence of the three species of seventeen-year cicadas: *Magicicada septendecim*, *M. septendecula*, and *M. cassini*. Cicadas emerge from the ground in great numbers at thirteen- or seventeen-year intervals over a large part of the eastern and midwestern United States, providing a feast for birds and mammals. We had never seen such a density of flying insects nor been deafened by any before. They would blunder into you or come down the chimney and squeal loudly when picked up. Rita Palin—a friend from Vancouver days—drove down from Ottawa to stay with us, and when she reached the center of town on Nassau Street, she pulled the car over and stopped, bewildered, convinced that the deafening noise of cicadas came from the siren of an out-of-sight police car behind her. For us, there was something special about experiencing them for the

FIGURE 13.4. Home, Riverside Drive, Princeton, New Jersey, in summer (**upper**) and winter (**lower**).

first time in Princeton in our first year, perhaps because they have the Princeton colors of orange and black.

In our early days in Princeton, Henry gave us invaluable advice on local and not-so-local natural history. The best places for walking locally were the woods of the Institute for Advanced Study and the towpath by Carnegie Lake toward the next village, Kingston. Farther afield, each requiring a two-hour drive, were the Delaware Water Gap for a hike through Mountain Laurel, and the Brigantine Marshes (Edwin B. Forsythe National Wildlife Refuge) for watching waterbirds. Visits to these two places became annual events. There were Black Skimmers to be marveled at in the waterways of Brigantine in summer and Snow Geese in winter. Henry introduced us to MacArthur's Bog, situated in the Greenwood conservation area of the Pine Barrens, at mile 20.5 along highway 539, leading toward the coast and the marshes. The bog lies on a hardpan soil with scattered quartzite pebbles and is surrounded by stunted junipers. Despite looking sterile, the bog supports some interesting plants including three insectivorous species: a pitcher plant, a sundew, and a bladderwort. Robert MacArthur used to bring ecology students here and posthumously and unofficially gave his name to the bog. With these field sites and other resources, I wanted to launch a field course in ecology and evolution, picking up where I had left off at McGill, but when Dan Rubenstein and I broached the subject with Henry, he was obviously upset at the prospect of playing second fiddle in his domain, so we dropped the idea.

A major reason for joining Princeton was the long-term stability, prominence, and collegiality of the population-biology group, so it was a major disappointment when John Terborgh left after our first year, and Bob May left after our second. Also, there were administrative changes afoot that resulted in me being appointed director of a new program, Ecology and Evolutionary Biology. The remainder of the Biology Department was cobbled together in a program called Cell and Molecular Biology. Arnie Levine was the director of that program but also the head of the newly structured Biology Department containing these two programs, so clearly the structure could not last long with such a huge imbalance of power. It was an interim measure, during which I had to make the case for a separate Ecology and Evolutionary Biology (EEB) department in a document with statistics to show how we would compare favorably with similar programs of the same size at other universities. Fortunately, Bill Bowen, the university president, was persuaded, and we were given the green light, with a muted warning that it was regarded as an experiment, and we had to prove ourselves. Well, we did, triumphantly. I regard the negotiated creation of the department as my best achievement as an administrator.

I learned of my appointment as director of the preceding program in the most unlikely of ways. We had returned to Genovesa Island in 1988. One of the friendly tour guides had brought a packet of mail, and not finding us in camp, he had placed the packet on a rock on the far side of the lagoon and then left. On returning to camp, we had already started cooling off in the lagoon when one of us noticed the packet, now precariously close to floating away or sinking. I swam over to retrieve it and on opening it found a letter from Dean Aaron Lemonick, addressed to all members of the Biology Department, announcing the new program structure, stating that I would be the director of EEB, and adding: "Peter Grant is in Galapagos on field research and I have not been able to contact him, but I am sure he will agree to the appointment."

When I was negotiating my appointment at Princeton, I secured an agreement to be off campus doing research on Galápagos for two to four months in the spring semester each year, because Galápagos was my lab, and spring was the time the finches bred. Permission to do this was a blessing for research but a burden for my roles first as director and then as chairman. "While the cat is away the mice do play," and while I was away some of my colleagues did play. I had support from the senior man of the department, the wise John Bonner, and from Dan, who followed me as chairman (for twenty-three years), but a couple of colleagues got up to mischief behind my back and made life difficult, just the sort of predicament that Mike Martin had alerted me to in Ann Arbor. Princeton, like Oxbridge, is prone to have cliques, and I was not present to control ours. In any event, I was relieved to hand the reins over to Dan, when he became full professor and chairman in 1991, and I took a well-earned leave of absence.

14

The Drama of El Niño

DARWIN'S FINCHES live in a dynamic environment. In typical years, a hot and wet season in the first four calendar months of the year is the time when they breed. There follows a cool and dry season, when breeding ceases and finding enough food to survive is at a premium. Every few years these contrasting conditions are amplified by the El Niño–Southern Oscillation phenomenon of extreme fluctuations in ocean and weather conditions, from cool surface water and droughts (La Niña) to warm water, high air temperatures, and abundant and heavy rains (El Niño). Extremes recur at about four- to seven-year intervals. The El Niño of 1983 was quite extraordinary. It was the most intense and prolonged event of the past four hundred years according to coral core data, lasted for eight months, and had profound effects upon the finches and all other components of the terrestrial ecosystem (Fig. 14.1). El Niño provided us with unique discoveries of competition, character displacement, and species formation.

The 1983 El Niño actually began at the end of November of the previous year with a rise in sea surface temperature and heavy downpours of rain. When we received the news of the weather shift from the Charles Darwin Research Station on Santa Cruz Island, Lisle Gibbs and his assistant, Jonathan Weiland, scrambled to get down to Galápagos as fast as they could. They arrived on Daphne in the latter half of December to find finches fledging from nests, and they did well to catch most of them. So began a seven-month field season under very difficult conditions with scarcely a break. We have enormous admiration for all they

FIGURE 14.1. Camping on Daphne Island, Galápagos. The only place to put a tent apart from in the crater is a small space on the outer slope. **Upper**: Wet season, 1975. **Lower**: Same view in 1983, an El Niño year, after abundant rain had promoted prolific growth and flowering of *Exodeconus* (*Cacabus*) *miersii*. (From Grant and Grant 2014, Fig. 4.10.)

managed to accomplish, and for two assistants on Genovesa for working almost as long and uninterruptedly. Together with our daughters, Rosemary and I completed the field season on Genovesa in August and then transferred to Daphne to do the same there. By the end of El Niño more than four feet of rain had fallen on this seasonally arid little island. We referred to the experience jokingly as a giant watering experiment that we summoned from on high to contrast with the drought of 1977. Jonathan Weiner exaggerated a little in referring to it as a flood (Weiner 1994). On the larger and higher islands, however, the relentless rains did indeed flood the land and create rivers, which makes sense of the reference to "pretty big Rivers" in the accounts of seventeenth-century pirates (Dampier 1927, 101).

In this extraordinary year, many finches bred when they were only three months old. This astonished us because we were convinced they could not possibly breed until at least a year old, like almost all other birds. The lesson learned was that some truths are revealed only under exceptional and rarely experienced conditions. A second striking event of that year was the establishment on Daphne of a breeding population of *Geospiza magnirostris* (Large Ground Finch). We had recorded vagrants in the dry season of previous years—most likely having flown over from neighboring Santa Cruz or Santiago Island—but they had disappeared when the rains arrived, probably returning to their island of origin to breed. Not so in 1982–83. Two females and three males stayed to breed and bred successfully. In subsequent years the population increased in size and was destined to have significant biological impact on the Daphne community of finches several years later (chapter 18).

The year 1983 was significant for a third reason; we were able to confirm the breeding of hybrids. From the first year of the breeding study (1976) onward, we had observed a small number of mixed pairs of breeding adults composed of resident *G. fortis* (Medium Ground Finch) and immigrant *G. fuliginosa* (Small Ground Finch) and, in one instance, a *G. fortis* × *G. scandens* (Common Cactus Finch) pair. None of the hybrid offspring survived to breed. We didn't know whether they were intrinsically weak or were dying through lack of food like so many other finches. I was predisposed to believe that Darwin's finch hybrids were intrinsically weak, having had experience of cross-breeding two species of *Clethrionomys* voles in the laboratory in connection with work on competition (chapter 10) and finding that the hybrids did not fare well—most failed to survive long enough to breed. Successful breeding of finch hybrids in 1983 showed us they were not intrinsically weak. In that year and the following breeding seasons, we discovered that hybrids would breed (backcross) with whichever species sang the song of the hybrids' fathers; if a hybrid's father was a *G. scandens* individual and the hybrid was a male, it sang a *scandens* song and bred with a *scandens* female, and if the hybrid was a female from the same family, it bred with a *scandens* male.

The occurrence of hybridization and backcrossing was as important a discovery as it was unexpected, because when we began a decade earlier, hybridization was presumed to be rare or absent among the finches (Lack 1947), and indeed among birds in general (Mayr 1963). Yet here on Daphne, it was clearly in evidence, with three species apparently exchanging genes. We saw hints of the same on Genovesa. Far from being an evolutionary dead end caused by weakness or sterility of the hybrids, hybridization, we could now see, potentially enhances the evolution of a species by introducing new genes (alleles). Moreover, it provided us with an explanation for why some populations of finches are so variable in beak morphology, and more than that, it gave us a new perspective on the necessary conditions for two species to coexist: ecological differences were not sufficient, mate choice differences were important too. Thus, hybridization emerged in our understanding as an important feature of the early stages of speciation in adaptive radiations; viability and fertility problems for hybrids evidently arise in later stages.

Beyond the world of finches, hybridization is now accepted by scientists as a widespread phenomenon. For example, in our own history, three human lineages exchanged genes about fifty thousand years ago: modern *Homo sapiens*, Neanderthals, and Denisovans. We contemporary humans have inherited the donated genes, as we know from genomic studies (Sankararaman et al. 2016).

And yet another significant consequence of the abundant rain in 1983 was a change in the composition of the seed supply to one dominated by small seeds. They were produced by vines, grasses, and small herbs that grew rampantly and smothered the low-growing *Tribulus* plants, whose large and hard woody fruits had been so important to finches in 1977. By itself the chánge had no immediate evolutionary effect on the finches, in contrast to strong ecological effects. All finches benefited from abundant food—mainly caterpillars—and bred with uniformly great success. Finches normally lay a clutch of two to four eggs, but some females laid clutches of six eggs and bred as many as eight times in 1983. Prolific breeding continued until the island started to become crowded, and breeding success declined. Two years later the island experienced another drought—only a single millimeter of rain was recorded in the rain-gauge—and once again many finches died. Now El Niño had its evolutionary effect. We predicted a slight shift toward a smaller average beak size caused by natural selection, and that is indeed what we observed. Because their preferred food of small seeds was relatively plentiful, *G. fortis* individuals with small beaks survived best, especially those with small and pointed beaks capable of picking up and rapidly processing a large number of very small seeds. Large seeds were scarce; hence more birds with large beaks died than those with small beaks, in striking contrast to events in 1977. The episode gave us a new perspective on selection over long periods of time, oscillating like a pendulum.

El Niño gave us the most important biological insights of the 1980s, and technology gave us important help in probing the genomes of finches. We learned from studies of other birds that the genetic signatures of individuals could be relatively easily determined from their DNA by sequencing regions of repeated units in the genome known as microsatellites. But would the method work in finches? To find out, we enlisted the enthusiastic help of our EEB faculty colleague Marty Kreitman. He discovered the method worked extremely well with dried chicken blood kindly supplied by a colleague from Rutgers University; the blood of birds has nucleated cells rich in DNA. Buoyed up by this, in 1988 we began to take a small drop of blood from the wing vein of every finch we caught, in a manner essentially the same as taking blood from newborn babies in hospitals for genetic typing. We transferred the blood to filter paper that had been treated with a binding agent (EDTA) to prevent deterioration, and then stored it in Drierite until it could be put into a freezer (set at −112°F) for long-term storage. Ken Petren, on a postdoctoral fellowship with us, identified unique markers that allowed him to characterize variation at sixteen genetic loci. These became a powerful tool for identifying paternity and maternity of the finches. Later they were repurposed for broader evolutionary questions concerning genetic differences between populations and the evolutionary history of the finches (phylogeny).

To our surprise, we discovered that the males we observed feeding young at the nest were not always their biological fathers, even though they were paired with the mothers. The biological fathers generally held territories close by. We had to adjust our estimates of lifetime reproductive success of individual finches to take this factor into account. No adjustment was necessary for females: the females we observed at a nest were the true biological mothers. Identifying paternity by genetic means was important for another reason: it provided confirmation of our observations of interbreeding and backcrossing. The genetic evidence enabled us to conclude that genes flowed from *G. fuliginosa* to *fortis*, from *fortis* to *scandens*, and to a lesser extent in the reverse direction, from *scandens* to *fortis*. Unexpectedly, a significant fraction of gene exchange, perhaps as much as half, occurred outside the pair bond, due to cryptic hybridization.

My release from chairmanship duties in the 1990s had brought about a turning point in our research in two respects. First, I was able to give more attention to applying to the National Science Foundation for funding. One of our previous proposals had been severely criticized by a laboratory geneticist who could not see the value of long-term field-work. Funding was minimal for three successive years, and much effort was required to get back on the NSF conveyor belt. Second, after one final field season (1991) when field asssitants stayed throughout the breeding season, we decided to handle the field research ourselves. This proved to be an invaluable return to everyday experience of the finches, but it came at the cost of documenting all the breeding, as we had done every year since 1976. We detected unusual things that our assistants had missed; we observed, for example, five male finches with exactly the same unusual song holding terrtitories next to each other. More than a decade later, this turned out to be a highly meaningful and important observation (chapter 18).

Our fieldwork was dominated by a preoccupation with the evolutionary importance of hybridization. Unlike short-term evolution caused by natural selection, hybridization is a long-term process spanning generations. We were finding that natural selection occurs occasionally—it is episodic—whereas hybridization is a more gradual and continuous process. By following the fates of banded birds year by year, we steadily accumulated profiles of their fitness—their survival and reproductive success. From the profiles we discovered that hybrids were remarkably fit; in fact, we could not detect a difference in fitness between hybrids and non-hybrids. Therefore, we concentrated on the genetic consequences of hybridization and the feeding characteristics and diets of hybrids as a possible explanation for their high fitness.

As on Genovesa, Rosemary and I shared and split the tasks between us, thereby ensuring that each could do the work of the other in the event of an injury (B. R. Grant MS). Measuring birds was in my domain; taking blood samples for DNA analysis was in Rosemary's (Fig. 14.2). We tape-recorded singing males and quantified the feeding of finches, either together or in different parts of the island. In 1992 we got off to a good start by catching the last two breeding birds of the previous year that had not been caught, and we measured and banded them. At this point, uniquely, every breeding finch on the island had been banded, which was a massive, long-term, cumulative achievement of our several students, assistants, and us, never to be repeated. Since 1991 had been an El Niño year, we thought quite reasonably there would not be another for several years. We were wrong. El Niño conditions continued for two more years, and in both of them we left the island too early to cover the full breeding season. Twenty years of Galápagos experience had not made us all-wise prophets of the weather.

Daphne experienced the full range of weather conditions in the decade of the 1990s, but there was no more selective mortality. Precisely

FIGURE 14.2. **Left**: Weighing a finch. **Upper right**: Weighing a finch, while Rosemary takes a blood sample from another finch, in a cave used for banding finches (photo K. T. Grant). **Lower right**: A cave used as a laboratory and kitchen. (All photos on Daphne, from Grant and Grant 2014, Fig. 2.13.)

two millimeters fell in the rain gauge in 1996, and for the first time since 1983, the island seemed to be reverting to the hard, bony appearance it had before the heavens opened: plenty of bare, exposed rock and a diminishing cover of dead herbs. Unlike those earlier days, however, the island now had a thriving population of Magnificent Frigatebirds, and in 1996 we confirmed the previously suspected breeding of the second species, the Great Frigatebird. The ecosystem, and not just the finches, continued to change.

The drought of 1996 was followed by an opposite extreme, an El Niño of extraordinary intensity that began in 1997 and became fully developed in 1998. It was actually hotter than in 1982–83, although with somewhat less rain—but still three feet (922 mm). The combination of heat and rain meant prolific plant growth on the island, and a smothering of cacti so great that when we counted fruits on our standard ten bushes in February, usually in the hundreds or thousands, there were no more than fifty-three! On the other hand, the crop of *Bursera* berries was one of the largest of all time. Finch breeding was in full swing, and with help from Ken Petren and Lukas Keller, we spent our whole time monitoring nests and building up an invaluable store of DNA samples and information on breeding success. Not everything went well. It was a bad year for blisters, probably caused by unusually abundant blister beetles that carry a Viagra-like sex-stimulant on their feet. Rosemary's legs were smothered in blisters. A couple of months later she had to have an ovariectomy, and we are convinced the two are connected.

On February 26 the sun almost disappeared. We wondered how this would affect the finches and positioned ourselves in the crater to take advantage of this exceptional event. I wrote in my notes: "95% eclipse of sun at 11 a.m. had little effect on the singing of birds in the crater. Sky went dark gray, a strange quality to the light reflected from the green vegetation. Light rapidly brightened after 11:05, but with noticeable effect upon the activities of the finches." The finches stopped flying, and this was expected, but they kept singing, and this was not. It was a mixture of day and night to them.

Our discoveries on Daphne and Genovesa reached a vastly greater audience than we had ever reached in our own publications when Jonathan Weiner published *The Beak of the Finch* in 1994. A superb book about evolution and Darwin, it treated us, our assistants, our research, and our family as a focus with a human interest to show how scientists work in the field and how contemporary evolutionary change on Daphne was discovered and interpreted. Jonathan writes like a dream, and the book was an enormous success that earned him a Pulitzer Prize. The only people who did not like it were the creationists. *Creation Research* magazine's criticism was actually quite mild: "Wonderful study, great data, wrong conclusions." Rosemary and I envied Jonathan's skill in weaving a compelling tale and wished we had the same talent. In the lead-up to publication, we had many meetings with him in Princeton and a very small number of disagreements and misinterpretations to iron out. We disagreed with him, for example, about what we said and how I expressed myself in explaining something to him. He was reluctant to give up his impressions because he quoted me liberally. Initially he portrayed me as more inarticulate and disjointed in my speech than I believe I am, and he completely mischaracterized Rosemary as playing a subordinate role to me. At one point he wrote, of Rosemary replying to me, "Yes, she said, sweetly." Nicola skewered that one. "Jonathan, you have to take that out. Mum never says anything sweetly."

Jonathan visited us for a few hours on Daphne, delightfully eager to see and find out about everything and ever quick to note down a salient fragment of our conversation. When he left, we waved and laughed at the same time. Understandably, he thought we were laughing at him, but actually we were laughing and commenting on the sharks circling the Zodiac! Several years later a friend of ours sent an article from the *New York Times* with an entertainingly ambiguous quote: "Weiner writes about the research of Peter and Rosemary Grant, who studied Darwin's finches on the island of Galapagos. . . . This book provides a wonderful argument for working hard to save them."

One thing led to another. Stimulated by reading *The Beak of the Finch*, Bill Kurtis came to Daphne in 1995 to interview us and make a film about the work. He stayed for one day only, but the prior planning, hard work, and excellent filming by naturalist-photographer Neil Rettig ensured they had plenty of material when they left. From our point of view, the filming went well, but we felt uneasy about how the script was going to be constructed, because the only biologist among them, Neil, was not going to have any say in it. There were initial problems of interpretation of natural selection and evolution, such as, for example, the layperson's misconception of adaptation as an intentional change for a purpose, as in, "finches increase in beak size to exploit the food supply better." The worst ones were straightened out in the final version, and the product was excellent: *What Darwin Never Saw* was a Public Broadcasting Service (PBS) success.

Three years later we were filmed by David Parer for the Australian Broadcasting Corporation. He misjudged our ability to be spontaneously eloquent in front of a camera in the brain-deadening atmosphere of El Niño. At one point, exasperated with my performance though beautifully under control, he said something like, "Now, Peter, that was fine, but just let's try it one more time, and this time imagine you are Ronald Reagan having one of his cozy fireside chats to the nation." There are a hundred and one smart replies to this—"Ronald Reagan is beyond my range," etc.—but it shows how numbing the heat was, that I couldn't think clearly and quickly enough to produce even one single response—rueful echoes of being tongue-tied in my Cambridge exam (chapter 5).

15

Uno Becomes Duo

LIVES CHANGE in middle age, and middle age for us was our fifties. For example, I started wearing glasses for reading when I was fifty-four. I remember the exasperation of wanting to look closely at a small insect on a leaf but not being able to focus on it as usual and having to back away, and by doing so almost losing sight of it! My ability to read numbers on metal bands on finch legs with the naked eye had begun to wane.

The 1990s were a decade of births, deaths, and marriages, also illnesses, but a time of greater freedom from administration, plenty of travel, exciting continuation of the Galápagos research, and a professional turning point for Rosemary. After I relinquished the chair of the department, we took a year's leave of absence in Europe. We had intended to spend the whole year at Uppsala, but it soon became clear that Rosemary's father's health was declining, so we spent the autumn and early winter of 1991 at his home in Arnside. I helped as much as I could by being a part-time amateur nurse and an elevator of spirits, while the others in the family did much more. He died at the end of November. Four years later Rosemary's mother died, at age eighty-nine. Both parents were in robust health until close to the end, apart from arthritis and age-related wear and tear.

My parents were not so fortunate. They were both heavy smokers. My father died at the tragically early age of fifty-eight, while I was in Montreal, and my mother died at eighty-four, both of lung cancer. My mother's downhill decline lasted for several months, was sometimes painful, and was made worse, not better, by a poor local National Health Service setup and hospital in Winchester, so my sisters told me. She returned home, and I visited her a few weeks before the end. She was calm, stoical, self-absorbed, and gave me the impression of being permanently bewildered by her fate. An intelligent woman, with a fondness for Roman and Greek history and mythology, she remained an enigma to me, as I am sure I was to her. Perhaps I reminded her too much of my father. She once said to my sister Sarah, while sitting in the audience before the ceremony at which I became a fellow of the Royal Society, "If Peter is so clever, why didn't he do something useful, like become a doctor or a lawyer?" Really. She thought I grew a beard to hide a weak chin, but a weak chin I do not have! Also, I supposedly never liked sports, which is a good example of the sixth of seven sins of memory (Schacter 2001), that of bias, a sin of misremembering the past so as to make it more consistent with current knowledge and *beliefs* (italics mine).

Both our daughters were married in the 1990s, first Thalia to Michael in 1992 and then Nicola to Ravi in 1994. We traveled to California for Thalia's wedding in beautiful October weather. Warblers danced above a fountain in the background as the resplendent bride and groom made their wedding vows in front of a priest from South Carolina. The marriage lasted only three years. We knew that Thalia and Michael had been raised in very different circumstances, but it seemed they were capable of surmounting any difficulties arising from their different cultures. On a visit to New Mexico in 1995, we enjoyed their company on a trip to see indigenous pueblos in a pine-filled, rocky canyon. The air was scented, light, and uplifting; it was sunny, yet the marriage, unknown to us, was severely fractured and broke soon after. Thalia's divorce was uncontested.

Nicola's wedding took place in our garden. We drove to a New Jersey town improbably called Berlin to pick up and rent a *mandap*, a wooden structure used for Hindu weddings, and then had fun erecting it in the garden at home. This was at the end of May, a little late for the azalea blossoms, just after Nicola and Ravi graduated from Brown University as medical doctors. We entertained a very large number of guests at home in the evening before the wedding day. We worried that our house would be too small for the guests but were reassured by the father of Divia, a close friend of Nicola, who told us that Sai Baba would ensure that everything would be all right. "You would be amazed," he said, "how the walls simply expand when the room becomes full." Earlier that day I had been widening columns in an Excel file with the cursor on my computer, and I could not help thinking of an all-wise Sai Baba—an Indian spiritual leader—using some divine equivalent to move the walls of our living room. That turned out to be unnecessary.

The Hindu wedding was a splendid, friendly, and colorful occasion. The priest helped the guests with English translations, gave guidance on when to take photographs, and issued occasional witty remarks. All went well until he started to read out the wedding vows for Nicola and Ravi to repeat—they were the wrong ones! Nicola and Ravi looked at each other and without speaking decided to go along with them.

Marriages were followed by births. First to be born was Nicola and Ravi's Rajul, in 1996. We threw ourselves into grandparenting. I experienced the same extraordinary feeling when cradling the newborn in my arms as I had when Nicola was born. We performed a lot of babysitting and learned the fallacy of the adage that grandparents "get all the benefits and none of the responsibility." Rosemary in particular was truly frightened at the prospect of something going wrong with Rajul when he was under our care, never to be forgiven by his mother. He turned out to be a fun-loving, athletic little boy—for example, serving at tennis when he was in diapers and no taller than the net. Thalia had hitched her wagon to Greg, a tour leader in Galápagos who we all knew, and Olivia was their first child. She was born the year after Rajul, in Galápagos, a Galapagueña. Our first impression of her was a quiet baby with dark eyes and a small mouth. A fraction of her DNA is Cherokee in origin, coming from Greg's mother. Anjali followed Rajul; Devon followed Olivia.

At the age of fourteen I broke my collarbone when playing hockey at school, but apart from that I had never broken a bone until 1993, when I tripped over an exposed pine root on the Appalachian Trail while walking and at the same timing looking at birds in the treetops. The result was a smashed and dislocated right shoulder. I had given a seminar at the University of Tennessee the day before and was lucky that our host, Stuart Pimm, was with us on the trail because his father-in-law was the senior surgeon at the Knoxville Hospital and arranged for another surgeon to operate. This excellent surgeon needed five hours to put the arm back in the socket of a broken shoulder after being forced to cut and then repair the elastic-like tendons holding it in the wrong place. Meanwhile, Rosemary had to endure a long stretch in a waiting room with two families of victims of a bloody fight who were now in police shackles. It was Saturday night.

After almost a week in the hospital, I had a long period of physiotherapy with experts in torture and spent a large amount of leisure time lifting cans of soup and attempting to push the dining room wall over, which was no easy task, as I had lost 80 percent of muscle mass in the upper arm during six weeks of immobility. For a long time, it looked as if the nerve damage was so severe that I would never be able to lift my arm more than fifteen degrees, but there came a magical moment when an electrophysiological shock elicited a weak response from all three branches of my axillary nerve. The neurologist, Dr. Frank Livingstone, shared the emotion with Rosemary and me; I even saw a tear in his eye.

In 1992 we stayed in the garret flat in Villa Åsen in Uppsala and worked there or in the Zoology Department. This visit was memorable for two activities. One was the start of occasional picnics in the late afternoon or evening with smoked fish (whitefish or trout), bread, fruit, a bottle of red wine, and coffee. We called them dead-fish walks (DFW) because the fish was smoked, and we have continued them every summer since. The first was an unambitious walk by the Fyris River to the edge of the town, to the fields full of reddish or yellow *Fritillaria* flowers, and to a convenient spot to sit down in seclusion and listen to snipe "drumming" on display flights and Thrush Nightingales singing as we drank our wine. On the last of the year's four picnics, we extended this excursion all the way to a lake, really an inlet of the sea. Being the middle of summer, it was still light when we returned at eleven p.m. The other activity was bird-watching. Staffan and Arne Lundberg took us out one evening to listen to Corn Crakes, Grasshopper Warblers, and a River Warbler. I saw the latter two more easily than I heard them owing to a bad fungal infection in both ears. I was literally almost standing on top of the Grasshopper Warbler, singing nonstop in a tangle of ground vegetation, before I could say I heard it. Corn Crakes remained out of sight.

Switzerland became another Sweden for us, for the generous hospitality of friends, and with the additional benefit of alpine flowers in spectacular mountain scenery. Foremost among the friends was Uli Reyer, a Staffan Ulfstrand equivalent in the Institute of Zoology at the University of Zürich. He invited us to give a seminar, and as a nice reward he took us to a place where, we were told, we would be certain to see a Wallcreeper. I have had the "certain-to-see" advice more than once without a sniff of success (European Badger, Andean Condor, etc.), but this time the advice was on target; after all, the Swiss are known for their precision. Exactly where we were told to look, we saw not one but two. This is one of those strange, exotic birds that have seemingly adapted to their environment well beyond the norm, like the Kakapo in New Zealand. They maneuvered themselves across sheer cliffs on either side of a narrow mountain road as they foraged for insects and spiders in small cracks and crevices. Inconspicuously dove-gray as they forage, in flight they suddenly reveal flashing patches of crimson against the white-flecked black background of their wings. The sudden display of colors must startle insects into flight and make them easy to capture.

Another significant trip was to Jeizinen, a small Swiss village above the Rhône valley. At five thousand feet elevation, the village was perfectly positioned for us to take long hikes up to the alpine meadows, and it became a favorite haunt to which we returned a half dozen times over the next ten years or so.

After such enjoyable visits to Sweden and Switzerland, we made it a cardinal policy to accept any invitation to Europe to attend meetings or conferences, or to give seminars, and to add one or more days at the location to explore the towns, art galleries, or countryside. Our visits included an international genetics conference at Birmingham, a two-week stay in Churchill College in Cambridge arranged by David Stern, a former graduate student in our department at Princeton, and several months at the Max Planck Institute in Germany, supported by an

Alexander von Humboldt senior-scientist award. At Cambridge I had the pleasure of working in the Balfour Library of the Department of Zoology, just as I had done as an undergraduate. I prepared a paper for a conference honoring Theodosius Dobzhansky, the eminent Russian American geneticist and evolutionary biologist, and wrote it later at Jeizinen. Cambridge was a new experience for Rosemary, whereas for me it was a trip down memory lane, so little had changed. New to both of us was the pleasant experience of living in the German countryside. We worked at the Max Planck Institute Department of Ornithology in Möggingen (Radolfzell) with like-minded ornithologists, thanks to our generous host, Peter Berthold. He shared our interest in hybridization of birds and measuring selection.

Later that summer (1996) we joined Staffan, his departmental colleague and fellow ornithologist Jan Ekman, and ten graduate students from Uppsala for an exciting trip to Africa. This was our first and long overdue visit to Africa and probably Staffan's twentieth. We spent three enriching weeks of animal watching and camping in Namibia and Botswana on a tour with a safari company that took us to the Gobabeb Desert Research Station and Etosha National Park in Namibia, and to Okavango, Maun, Moremi, Savuti, and Chobe in Botswana. We left the group at Victoria Falls in Zimbabwe, flew to South Africa, and spent a day with Alan Kemp at the Transvaal Museum in Pretoria before returning to Germany.

A continent becomes a collection of vivid experiences: the light, air, space, and biological fascination of the first two countries we visited mixed with the subliminal threat of violence in South Africa. The big animals were exciting to see, large and approachable as I had expected. Unexpected was the thrill of seeing *Welwitschia* plants in the national park of that name. Could any plant be more improbable than this cone-bearing tree with its trunk beneath the soil and only two long, strap-shaped leaves lying in a tangled and twisted heap on the desert surface? It is so extraordinarily different from anything else in the world and so ancient, it encourages the mind to fly backward for a hundred million years or more and imagine dinosaurs and other evolutionary experiments in the desertscape. To say we could have stayed there for days is not very meaningful; we could have stayed for days at every place we visited—outside the towns, that is.

Lions, cheetahs, elephants, zebras, wildebeests—the list of magnificent mammals goes on and on. For the record, we saw more than forty species of mammals and 240 species of birds—a surfeit of riches. This is a world run by predators, unlike Galápagos. The only place where we could walk in the northern parks was in Okavango, because we had a registered guide. Map Ives, son of a Yorkshireman, born in Botswana, man of the bush and largely self-taught, answered all our questions and told us much more. He did not set out to scare us, but his various remarks about the need for caution and what to do in an emergency such as a lion charging drove home the point that we were in a potentially dangerous place, more dangerous than any other we had ever experienced. Getting almost within spitting distance of a spitting cobra did the same. So did the sight of an enormously thick, fifteen-foot-long crocodile suddenly rising out of the water and thrashing about in an attempt to separate the antlered head from the body of an antelope in its massive jaws. So did the lion that snarled at us when we approached the giraffe it was eating, and so did another that stared at *me* straight in the eyes at a distance of one hundred feet. Oh, those lowered yellow eyes! Ugh! Somebody said: "I have never been anywhere before where I felt part of the ecosystem, and not as the top predator!"

The mid-1990s were a turning point for us professionally. Although Rosemary and I had worked together on the Galápagos since 1973, the research was primarily my responsibility. She had taken a back seat in public for twenty years, and invitations to present our work at meetings came to me. For example, at the American Ornithologists' Union meeting in Minneapolis (1997), I gave a lecture to mark the fifty-year anniversary of David Lack's book *Darwin's Finches* (Lack 1947). In the same year I contributed a lecture to a conference in Irvine, California, celebrating the sixtieth anniversary of Dobzhansky's *Genetics and the Origin of Species* (1937). The transition to sharing the limelight took place over a short span of a few years. Rosemary entered the public limelight on an equal footing with me in three significant steps. The first was a seminar she gave at the University of Minnesota. She was nervous; nevertheless, she did well, although her lack of experience showed, and she did not feel good about it. I encouraged her to try again at the next opportunity, to get back on the horse that had just thrown her.

The second public speaking, in contrast, was a huge success. She was invited to a conference in Asilomar, California, in honor of Guy Bush, a distinguished and influential evolutionary biologist (B. R. Grant MS). In that setting, and in my absence, she performed excellently, by her account and those of others. The third was at the International Ornithological Congress in Vienna. The meeting had been dogged by various signs of disorganization, and what followed in her contribution to a symposium was just one episode of many. Right at the beginning, the projector jammed. The student projectionist lifted the carousel out of its position and turned it upside down to see what the problem was, and the whole set of forty slides fell onto the table in a heap. Rosemary had rehearsed her talk extremely thoroughly and proceeded to deliver it without notes or slides and with a happy smile on her face. She emerged from the ordeal with a reputation magnified many times and a large number of new admirers. I wish I had been in the room.

Uno (Ego) had become duo. Now the invitations came to both of us to share a lecture: in Sweden, in Switzerland, in Honolulu, and in the Hawai'i Volcanoes National Park on the island of Hawaii. Within a couple of years, Rosemary was advising me on how to improve my half of our joint talks (Fig. 15.1). Honors came to both of us as well. We jointly received the Leidy medal from the Academy of Natural Sciences in Philadelphia in 1994, which was a great honor because Rosemary was the first female recipient since the award was first bestowed in 1923. We jointly received the first E. O. Wilson Naturalist Award from the American Society of Naturalists in Vancouver in 1998 and were both elected members of the American Academy of Arts and Sciences in 1997. An annual lecture in our names was inaugurated at the University of Zürich several years later (2011).

The visit to Vancouver to receive the Wilson prize was most enjoyable because it was a return to UBC and an opportunity to visit favorite places in the vicinity like Mount Garibaldi and Mandarte Island. On a

FIGURE 15.1. **Upper**: In my Princeton University office in Eno Hall with Rosemary, November 2015 (photo Denise Applewhite). **Lower**: Discussions in my office, March 2014 (photo Ricardo Barros).

brief visit to Salt Spring Island, we had a rendezvous with two fellow graduates from days at UBC, Bristol Foster and Frank Tompa. In fact, this reunion was so enjoyable we began to think of a return to buy a piece of land.

In the year leading up to the next century, I was president of the American Society of Naturalists. Although I had the flu at the main meeting in Madison, Wisconsin, I enjoyed running the meeting, but with deafness making it increasingly difficult to understand several things said around the table, I decided it would probably be my last such participation as a society officer. My presidential address was uncharacteristically strong in tone and urgent about our environmental predicament (Grant 2000).

In December we had the pleasure of seeing the largest moon we will ever see, supposedly 14 percent larger than normal. It was not on our minds when, at about five p.m., we turned from Washington Road onto Prospect Avenue in Princeton, looked down the street in complete darkness, and saw dead center a large yellow ball rising above the trees. It was like a huge gold medal, awarded to humanity. For what? Surely not for taking good care of planet earth; our transition from tenant to custodian has hardly been a medal-earning success. As if to make the point, the medal shrank as it climbed in the sky, out of reach, beyond our grasp. The year, decade, century, and millennium closed with gloomy predictions of apocalyptic collapse, the so-called Y2K problem, with computers crashing, planes in the air without contact with air-traffic control, civilization crippled, and so forth. We stayed up to see in the new year/century/millennium. And nothing happened.

16

In Search of DNA

THE GUIDES had a joke for the tourists: once you have seen one island you have seen the lot. This is an outlandish claim because the truth is the opposite. "The most astonishing thing about the various islands of the Galápagos is their superficial similarity and their actual diversity" wrote American naturalist and explorer William Beebe (1924, 259). Differences among islands are reflected in the finches and are an important reason for their diversity.

Once we had a method for collecting and storing small samples of blood for later DNA-extraction and analysis, we had the freedom to expand the scope of research beyond Daphne and to ask new and far-reaching questions. The most important questions were about finch history: Where had the finches originated, when did they arrive in Galápagos, and how are the species related to each other? Which are the oldest and which are the youngest? DNA can provide answers because the history of a species is recorded in an individual's genes. Another question is this: How is the genetic makeup of the Daphne finches affected by natural selection and the propensity of finches to hybridize? And yet another concerns the status of small and potentially endangered populations. Finally, in the absence of fossils, DNA offers the best hope for detecting the signature of species that might have once existed but became extinct, like ghosts in the genomes of today's finches. We are still hoping.

While continuing the work on Daphne every year, we visited all major islands except Pinzón. At each site we captured birds in nets, as on Daphne; Rosemary took blood samples from every bird, and I

FIGURE 16.1. The family on Española Island, Galápagos, August 1980.

weighed and measured them, unless we ran out of time. This chapter describes the fun and failures of visiting islands in search of finch DNA.

We had been to Champion and Gardner Islands (both by Floreana), Floreana itself, and Española in 1979 and 1980 (Fig. 16.1) and returned to find the vegetation completely transformed by the rampant growth caused by the 1982–83 El Niño event. The same transformation had taken place on Genovesa and Pinta. We lamented the passing of the "good old days" when walking was not hindered by shrubby thickets of Chala (*Croton scouleri*) and Muyuyo (*Cordia lutea*). The scientific message was important: large changes in the vegetation can persist for many years—perhaps decades—as a result of a single extreme climatic event. Seeing these enduring effects changed our perspective on finch evolution and made it clearer than ever that a short field season in an environment like Galápagos has limited value when trying to understand long-term evolution. How much we would have missed had we stopped after three, six, nine, or twelve years!

At times we received as much if not more logistical support from the Galápagos National Park people as from the Charles Darwin Research Station personnel. The new director of CDRS in 1993, Chantal Blanton, was facing legal action for laying off workers. We took a small trip in February to La Bomba on the northeast coast of Santiago Island in search of *Geospiza scandens* (Common Cactus Finch), courtesy of the national park's boat *Lancha No. 5*. Museum collections had few specimens from this island, and we feared the population might be on the brink of extinction, as we had not seen any *G. scandens* on a similar trip to Buccaneer Cove a couple of years earlier. We needn't have worried. We found about a dozen finches where the crew dropped us at a park guides' camp, tape-recorded several, and caught four of them. The trip back in the rain was something else again, as I lay on rain-soaked newspaper on the smaller half of a triangular board occupied by the skipper of ample frame ("mi amigo"). I do not know how I slept or why I didn't roll off into the bottom of the boat, where a small saline and slightly oily lake was waiting.

The next trip was to Caleta Negra on Isabela Island to look for Mangrove Finches (*Camarhynchus heliobates*) in the big stand of Black Mangroves and White Mangroves, as there was little information on the status of this, the rarest of the finch species. We used playback of tape-recorded song to bring the finches close and saw nine. We resolved to come back and census the whole population here sometime, because nine was a worryingly small number.

At Punta Espinosa on Fernandina Island, we spent a day and a half looking for Mangrove Finches yet failed to find or hear a single individual. Apparently, they used to breed there, but the small population must have become extinct. The rest of the time we spent helping my graduate student, Galápagos-born Carlos Valle, in his study of the strangely inspiring Flightless Cormorant, photographing the cormorants, penguins, and much else. Seeing the cormorants again was a thrill (Fig. 16.2). They had made a big impression on me when I learned about them at school in an evolution lesson, and again when I saw them for the first time on a brief visit in 1983. Their sapphire eyes are as exquisite as they are rare in the world of birds. Does the color help them when feeding underwater? Fernandina feels remote, the more so when the

FIGURE 16.2. Flightless Cormorant (*Nannopterum harrisi*), Fernandina Island, Galápagos (Photo K. T. Grant).

fogs swirl in, as they often do in late afternoon and overnight, bringing a damp chill, splintering the sunlight and revealing numerous spider-webs by covering them in glistening dew. The mountain rim beckoned, but the worst clinker lava we had ever experienced said no.

The following year we ventured farther afield in formidable seas to Darwin, the northernmost island. We could not land in the preferred place beneath the talus slope; instead we managed to scramble into a cave high up on a sheer cliff when the *panga* (Zodiac) reached the crest of the swell. Rosemary scratched herself badly on barnacles and developed a strong allergic reaction that kept us in the cave all day. However, our three companions managed to creep along a terrifyingly narrow ledge above a vertical cliff, climb up, tape-record two of the extremely rare Large Ground Finches (*G. magnirostris*), capture Northern Sharp-beaked Ground Finches (*G. septentrionalis*; formerly *G. difficilis*)—our reason for coming—and then lower them in individual bags by rope at the end of the day for us to measure. Around six o'clock, our companions came down and joined us, and by the time we were ready to leave the island, night was fast approaching. The sea was rougher than ever, so we had to strip off most of our clothes, bundle them into garbage bags, and throw them and our packs into the waiting arms of a crewman in the *panga* below. Mercifully, not an item missed the boat. Then we dived or jumped off one at a time and swam to the *panga*. I was oblivious of the seven-foot hammerhead sharks down there as I made my way slowly back up to the surface, enjoying the warmth of the water, at 82°F. Rosemary followed. Each of us caused our waiting colleagues some concern when we did not pop up from the depths like corks. As I swam back to the *panga*, the thought occurred to me that Rosemary and I were a bit old for this sort of thing.

We failed to get on to the island the next day, and then failed to land at Elizabeth Bay on Isabela Island, where Mangrove Finches had once occurred. We failed to land at Cabo Douglas, Fernandina, for an intended climb to the crater's rim, because the swell was too great again, but in this case our failure was a success of sorts because on that day Fernandina erupted lava from a fissure high up on the outer slope of the western wall. The steam, smoke, and fumes would have made being there unpleasant and unhealthful if not actually dangerous. We did spend a day at Playa Tortuga Negra, on Isabela Island, assessed the status of the Mangrove Finch there, and later wrote a paper stressing the dangerously small size of the population (Grant and Grant 1997). The paper stimulated a long-term and continuing effort to conserve the species sponsored by the Durrell Wildlife Conservation Trust. Numbers are now known to be ten times higher than we believed, but the number of breeding pairs is closer to our estimated small number. The need to conserve the species has since been strongly reinforced by the discovery of persecution by rats and a fly (*Philornis*) whose larvae eat and sometimes kill finch nestlings.

The day after Fernandina erupted, "hostilities" erupted between Ecuador and Peru in the jungle borderland, but fortunately, no more than threats of invasion reached Galápagos.

A relative of *Geospiza septentrionalis*, the target of our trip to Darwin Island, is present on Santiago (*G. difficilis*), so Santiago was the next island on the agenda. We joined our Princeton colleagues Martin

Wikelski and Michaela Hau, and with the aid of national park personnel and two burros, we climbed to the top of the island. As we passed through the lowland forest, we wondered whether Darwin had seen the largest of the *Bursera* torchwood trees. This safari-like enterprise was a new experience for us in several ways—sleeping and cooking in a hut in the highlands, having a shower while surrounded by a pastoral land-scape, and so on. We thoroughly enjoyed it, undoubtedly because the weather was conveniently dry, and because we were successful in tape-recording and capturing about a dozen *G. difficilis* individuals for blood samples. We were there for only a day and a half, including two nights, but long enough to do a little walking around and exploring. Large, soli-tary tortoises stood motionless nearby, and momentarily we were able to shut out the devastation to parts of the habitat caused by goats and imagine an earlier, even pre-Darwin state. The calling of Hawaiian Pe-trels early in the morning added to the sense of wildness.

The pursuit of *G. difficilis* led us to Fernandina, and this time we had planned the trip well. Simón Villamar from the national park and two assistants had already taken water and food to a camp on the rim. On the long climb up, we made two lengthy stops at well-chosen shady places. Cloud cover protected us from the strong sun, and by the time we reached the rim we were far from exhausted. We put our packs down, walked to the rim on black, sand-like granules of lava, past a few low bushes, and gazed in amazement across the vast caldera, a huge hole in the landscape. It was the sort of view that inspires awe on the rim of the Grand Canyon. Steam was rising from a couple of places, and three small pools could be seen amid the tangle of lava at the floor. I joked with Rosemary that, at sixty, she was the oldest grandmother to climb the island with a pack on her back. After putting up our tent and eating lunch, we decided to work. As it turned out, the afternoon was bonus time for netting, although our work was truncated by heavy rain. The next day we made a tactical error and, instead of netting close by, chose to walk around the rim westward to the point straight across from our camp, where we could see a forest of *Zanthoxylum* (cat's-claw) trees through the binoculars. This was a fascinating day, although useless from the point of view of our scientific objectives. We passed fumaroles, climbed down on to an upper platform inside the caldera, and photographed Galápagos Land Iguanas. Heavy downpours precluded netting and restricted it to three hours the next morning. There was no problem in getting down to the coast the following day, other than avoiding the slippery rocks.

Later that year (1997) we joined a Smithsonian research vessel, the *Urraca*, at Barro Colorado Island in Panama, and three days later arrived at Cocos Island, home of another Darwin's finch species, *Pinaroloxias inornata*, the Cocos Finch. We spent two weeks on the island with Bill Eberhard, a specialist in spider biology from the Smithsonian, and stayed in a splendid Danish-designed building still being constructed by the Costa Rican National Parks guards/carpenters. They were our companions for meals—delicious black beans and rice, at every meal. Prepared to camp, we were fortunate to have a waterpoof room because rain fell heavily on most afternoons. Capturing finches around the kitchen and near the pigpens was very easy. We captured thirty-five birds in two mornings and then made things difficult by netting across the Río Genio in more natural vegetation. The finches were incongruous. They looked so obviously like Darwin's finches, but here they were in a rain forest (Fig. 16.3). When not recording their feeding habits, we explored much of the island, either soaked by rain or baked by sun.

I have strong memories of this rain-forest island in the middle of nowhere: the finches, flycatchers, and cuckoos; the tall trees, especially *Cecropia* and palms; the strangling *Clusia*, with extraordinarily thick, dark-green, platelike leaves and roots that start at the *top* of the host tree; sunlight through the epiphytes; the slippery roots and rocks at my feet and slippery rocks in the swollen Río Genio, which we navigated perilously with poles; the numerous migrant species of birds, and the dead migrant martins on the ground in the morning; the two "wild" cats around our building; rats and pigs observed in the forest; the large blue butterfly flying across the river; little striped *Anolis* lizards; the first sight of the electric-blue arms of a crayfish in our local stream; and the afternoons with feet up on the veranda, sipping Camping Gaz–cooked coffee and reading Janet Browne's *Charles Darwin* (Browne 1995) while the

FIGURE 16.3. Rain forest on Cocos Island, Costa Rica, 1997. (From Grant and Grant 2008, Fig. 6.)

rain came bucketing down all around. Fernandina and Cocos explored in one year, two major achievements!

Our last Galápagos field season of the 1990s was marred by a theft of many of our personal possessions and camping supplies. We had spent three dry weeks on Daphne, then we went to Isabela Island and climbed Volcán Alcedo for the first time in search of *G. difficilis*. We did not find it, but searching for it and trapping Green Warbler-Finches (*Certhidea olivacea*) and other finches was fun. We stayed with a national park guard on the rim in a hut that gave us dry comfort and a stove to cook on. The view across the caldera was spectacular. Sometimes the crater was full of cloud while the sky was clear above; at other times it was the reverse, and two tall columns of steam arose from fumaroles to the west of us. Most memorable were sunsets, in which pillars of clouds rising above the caldera caught fire in the rays of the setting sun, reminiscent of paintings in the rococo style by the Italian artist Tiepolo, only without the floating goddesses! I wish we had been able to hike around the rim for more than the quarter segment we accomplished. Destruction of habitat by goats was sad to see; there were fifty thousand of them on the island, according to one estimate. Nevertheless, there was enough vegetation around and below the rim, and an abundance of lumbering, grunting, and occasionally mating tortoises, to give us an overall impression and sporadic glimpses of wilderness. It was exhilarating.*

The borrowed packs we were using did not fit very well, and we practically collapsed on the ground upon reaching the coastal camp and taking them off. I walked two hundred yards to the overhanging tree where we had left our personal belongings and entered a nightmare. I could not believe my eyes. Our day packs were there, and most of my clothes were strewn on the ground. Gone were the main bags. Gone were Rosemary's clothes, personal belongings, spare electronic equipment, many things from our stay on Daphne, including blood samples, and gone was our tent. It is a virtual certainty that the culprits were from the fishing boat that had been hanging around when we arrived nine days before, appropriately named the *Corsario Negro*. Carlos Valle had previously caught them redhanded stealing from his camp on Fernandina.

We were in something of a daze for the rest of the afternoon, eating to replenish lost salt and energy, washing some clothes and hanging them to dry on the bushes, and putting up a makeshift lean-to tent with mist-net poles and a tarp we had taken to the top. That night was one of the worst ever. When it was not blowing a gale of rain through our "tent,"

---

* Galápagos National Park guards eliminated *ALL* the goats later!

threatening to lift it up and carry it away, the mosquitoes scoffed at our insect repellent and drank our blood. This went on all night. When the pain of one bite began to wane, another started with fresh intensity. Our space blankets were a godsend, but even they could not protect us. Next day, worn out and walking around like zombies, we tried to sleep but failed. We did sleep solidly that night; not even mosquito bites kept us awake. The following morning the *Beagle III* arrived, rescued us, and so ended our field season.

The year 2001 marked the last of our long field seasons. We were intrigued by the fact that the two oldest species at the base of the Darwin's finches phylogenetic tree—the warbler-finches, *Certhidea olivacea* and *C. fusca*—occupy different islands and never occur together. In combination they occur on all major islands and several minor ones, so they surely have had plenty of time and opportunities to become joint occupants of an island but have not done so. Why not? Some individuals of one species must have flown to an island occupied by the other at some time in the past yet failed to establish a breeding population. Either they did not differ enough ecologically to coexist, and one competitively excluded the other; or they did not differ enough in courtship signals and responses, interbred, and by backcrossing to the resident species, the invader became "reproductively absorbed" (Grant 1986) into the resident population. On Daphne we had discovered a clear example of the second possibility in *G. fuliginosa*. Widespread elsewhere in the archipelago, this species has failed to become established as a separate population on Daphne because the rare immigrants breed with the much more numerous *G. fortis* (chapter 14). They are not competitively excluded by *G. fortis*—their food supply, principally small seeds, is plentiful, as it is also on islands both smaller and larger than Daphne.

Would *C. olivacea* and *C. fusca* interbreed, we wondered, if they ever encountered each other? To test this indirectly, we played the tape-recorded song of one species to the other species, and measured reactions of residents to the pseudo-interloper. The tests were accomplished smoothly and quickly on Santa Cruz (*C. olivacea*) and on Genovesa (*C. fusca*), but by the time we reached Pinta, breeding was almost over, and the birds' motivation levels had plummeted. Even with incomplete results, the mutual responses we found were striking, and they enabled us to conclude that females would probably recognize males of both species as potential mates. Perhaps interbreeding does happen occasionally, and stray immigrants simply get absorbed into the receiving populations. Alternatively, the hybrids may be intrinsically less fit, and finally, we cannot rule out the possibility of competitive exclusion, although this seems to us unlikely.

We camped comfortably behind the sea lion beach on Pinta in the company of our friend Tjitte de Vries, now a professor at the Pontifical Catholic University of Ecuador in Quito. Our trip was marred, however, by some fishermen—possibly from the *Corsario Negro*—who we believed intended to steal some of our belongings at camp. On our first full day, as we were walking inland, three *pangas* made a sudden dash to our camp from the entrance to the bay, so we hurried back to the camp and arrived just in time to forestall the visitors. They pretended to be innocently cleaning their engines. A standoff lasted for twenty minutes, and then they left. Two days later the performance was repeated, one hour earlier, and the following day it happened again, one hour earlier still, but we were in the vicinity, and the fishermen did not even come ashore. After that we alternated standing camp defense with Tjitte, whose study of hawks was unrewarding since the birds were not breeding. Our time was not wasted on our days near camp, as we tape-recorded a large number of warbler-finches. The threat from the fishermen disappeared when a large oceanographic research vessel arrived and left two people on shore for our last two days.

We completed our archipelago-wide survey with trips in 2003 to Rábida, Floreana, and San Cristóbal Islands, all for purposes centered on DNA and song. The Rábida visit began with the discovery of a picturesque campsite just beyond the tourist trail. Sunrise was spectacular, viewed as an orb flanked by two silhouetted *Opuntia* cactus bushes, their flat and prickly pads extending sideways, seemingly stuck on at odd angles. After the almost daily group of tourists had left, before half past four, we would go down to the White Mangroves, where we stored food and water, and wash and watch penguins and pink flamingoes in the lagoon. This was the first time we had seen flamingoes up close, and we could observe their feeding behavior and strange locomotion accompanied every now and again by barnyard-goose-like honking. When they took off in flight, they seemed unable to take a democratic decision on where to fly, like a committee that has lost its chairperson.

We had hoped to capture and tape-record many finches, but the island was as dry as a biscuit, and we had to settle for about a dozen captures and a few tape-recorded songs. The highlight (my hip disagreed) was climbing to the gritty top of the island. This entailed negotiating bands of bouldery rocks and *Croton-Bursera-Cordia* thickets. None of the forty-nine stable rocks prepared us for the treacherous fiftieth. At the top were small clusters of *Macraea* plants, looking at a distance like conifer saplings. We saw only one *Certhidea* individual on that climb and counted ourselves lucky to have caught three out of a total of five seen during our stay. Coming down the island took precisely five minutes less than going up. We were tired and more than usually careful.

San Cristóbal is a large inhabited island in the southeast, the oldest and the first that Darwin visited. We journeyed overnight to get there, rented a taxi, and drove to the highlands to tape-record warbler-finches in order to compare their songs with songs on other islands. The finches were plentiful, and in about two hours we accomplished our goal of tape-recording more than thirty individuals. In many places an extraordinarily tangled growth of the introduced blackberry (*Rubus*) formed an impenetrable barrier to the forest. Having gained spare time, we spent the afternoon at the lake El Junco and at the Galapaguera tortoise reserve. It was carnival day and night in town. On board the *Beagle III*, anchored at Puerto Baquerizo Moreno, we got little sleep, enveloped in the singing and music broadcast from the quay at full volume, until our boat left at four in the morning.

The following year we completely failed to find *Certhidea* in the highlands of Floreana, despite help, advice, and companionship of the landowner Claudio Cruz, and wrote a paper drawing attention to its probable extinction "behind our backs," unnoticed and unrecorded.

17

First Experiences of Asia

I FEEL MOST English when I'm listening to carols at Christmas, recorded at Oxford or Cambridge, reminding me of childhood and later listening to them in Great Saint Mary's Church in Cambridge. For example, at the start of 2001, we celebrated the end of skiing, the day before wet snow ended it, with mulled wine by the fire at lunchtime listening to Christmas music. On the other hand, in American society I am rarely aware of being English, although once in a while I cannot avoid it. On Rosemary's sixty-fifth birthday, we listened to an evening lecture by the widely known ecologist and conservationist Paul Ehrlich in McCosh Hall on the Princeton campus, which was preceded by this amusing vignette. Rosemary and I sat down in the front row only a few minutes before the lecture was due to start. An old man in the row behind us, leaned forward, tapped me on the shoulder, and said, earnestly, "You look like Charles Darwin," "Oh, do I?" I said. "That must be because I am wearing an English sports jacket." "Yes," said he, ignoring my remark and lowering his voice ominously, "just before he died." Well, everyone cracked up with laughter, but he appeared not to notice, as he continued in even lower register: "He was seventy." Fortunately, the man was no prophet.

Our lives revolved around activities in three locations: Galápagos, Princeton, and Elsewhere. In Princeton we analyzed the data we brought back from Galápagos and wrote papers for scientific publications. In 2008 we published a small book, entitled *How and Why Species Multiply*, that was designed to embed our Galápagos findings in the general context of explaining how species came into existence and persisted (Grant and Grant 2008). "Elsewhere" was the many places from around the world where we were invited to give seminars on the Galápagos findings. Such visits have enabled us to explore the surrounding country or urban attractions. This chapter is located mainly in the Elsewhere zone, and touches on the highlights of those experiences.

Domestic travel took us to the American Institute of Biological Sciences' meeting in Arlington, Virginia; to Saint Louis to celebrate the fiftieth anniversary of the Missouri Botanical Garden; to the Rocky Mountain Biological Laboratory to participate in its seventy-fifth anniversary, and to the thirty-ninth Nobel Conference held at Gustavus Adolphus College in Saint Peter, Minnesota. This last conference, with the theme "The Story of Life," was a unique experience for us as it was held in an ice hockey arena. The audience comprised mainly college and high school students and teachers, forty-five hundred of them, and another fifteen hundred watched on screens in overflow rooms. Everything about this conference was superb, from the moment we arrived and were greeted with Swedish-style hospitality. There were excellent talks, several musical interludes, and one musical evening. Rosemary and I had about eight questions to answer after our talk, and before we left the podium, I gathered up thirty-one pieces of paper, on which a total of forty questions had been written and submitted to the chairman. Some were so simple, direct, and penetrating, they made us say, "Why didn't we think of it that way?"

In 2002 Rosemary was invited by Uli Reyer to be a visiting professor in the Institute of Zoology at Zürich for six months, so we arranged to spend half of two successive summers there (Fig. 17.1). Rosemary's task was to teach all comers in a seminar course on speciation and hybridization, and I was the teaching assistant, back in the role I had had as a graduate student. We installed ourselves in a student apartment complex known as The Network, on Bülachstrasse, a mere ten-minute walk from the department, where our office was luxuriously spacious. It overlooked a serin's nest in a pine tree, and on a clear day we could see the

FIGURE 17.1. With Rosemary, a visiting professor at the University of Zürich, Switzerland, 2003. (Photo Rüdiger Wehner).

snow-capped Alps in the background. We loved the spring weather, walks up to Zürichberg, listening to blackcaps on our way home to lunch, and having dead-fish suppers on campus on sunny evenings.

The pace of life was so gentle in comparison with Princeton. People had time to stop and talk, and we fell into line with local practices. The city center was a ten-minute tram ride away. We took advantage of exhibitions of the works of J. M. W. Turner and Alberto Giacometti, watched *La Traviata*, and listened to an amazing solo evening performance by Elena Mosuc in Lucerne. Her voice was incredibly powerful yet controlled, and beautiful, quite the best soprano we have heard since Maria Callas.

And then we spread our wings and flew farther than ever before. Jae Choe had invited us to the eighth INTECOL (International Association for Ecology) meeting in Seoul, South Korea in 2002. Rosemary was to give a forty-minute pre-banquet talk on what it is like to work on the Galápagos, and I was to give the keynote address the following morning. Neither of us will ever again be heralded to the stage with a fanfare of trumpets, I thought, or make polite conversation with a mayor and the governor of a province (I was wrong), and in my case, never again will I

be thanked by the president of a country (Kim Dae-Jung) in a prere-
corded video welcoming us to the meeting. He made impressive and
wise remarks on why we need to respect the environment.

We were to fly to Beijing in the late afternoon after my talk in order to
attend the International Ornithological Congress (IOC). To compen-
sate for this unfortunately early departure, we arrived in Seoul a few days
early. Jae gave us five student companions and a rental van. They took
us to the Changdeokgung Palace, the Korea National Arboretum (out of
town), a Korean folk village, and finally to a traditional music concert at
the Seoul Arts Center, where we were enchanted by a monk's dance and
extraordinary drum playing. At the INTECOL banquet, we listened to
six- and twelve-stringed zither music with unusual rhythms and an extraordinarily beautiful, haunting piece of music, called *Road to Sampo*,
played by a young woman on a *daegeum*, a type of cross flute. This was a
new type of popular traditional music (*Gugak-Gayo*), reminiscent of
musical sounds from the Andes. Better than the banquet itself was a
lunch we had with the students. Seated around a barbeque grill in the
middle of the table, we cooked and ate thin strips of beef, decorated with
vegetables from various small dishes, enhanced by spices, and wrapped
in sesame leaves. Delicious! Later, in a museum in Xi'an, China, we were
to see a barbeque grill of almost the same type made a couple of millenia
beforehand. We left South Korea with memories of the grace and gentle charm of our various hosts; the many types of singing cicadas, and
one father and son in a park catching them to put into a small bamboo
cage; the extraordinary density of Lego-like high-rise apartments
from one end of the horizon to the other; horrendous traffic; wealth
and Western-style prosperity, for some; enormous billboards adver-
tising weddings, with bride and groom looking very Western; tasty,
spicy food.

Rosemary's plenary address at the IOC gained her prolonged applause,
a standing ovation from part of the audience, and a plethora of accolades
over the next two weeks. Outside in Beijing, the atmospheric pollution
was the worst we had ever seen. I doubt if we ever saw anything more
than a half mile away. It was a pleasure and a relief therefore to leave the

city and embark on a tour to Lhasa, Tibet, via Xi'an and Chengdu. The
first stop was the Great Wall, where we had to run the gauntlet of potential pickpockets and a thousand vendors ("Hello, one dollar!") to
enjoy the relative freedom of walking along/climbing a few hundred
yards of the wall—enough to imagine the remaining thousands of miles
built more than two thousand years before. The second stop was the
tomb of the terra-cotta warriors and horses at the Mausoleum of the first
Qin Emperor World Heritage Site and the Banpo Museum, in Xi'an. The
warriors were discovered by a farmer in 1974, two years before Mao
Zedong died. They are an amazing, not to say fantastic, spectacle, as well
as a testimony to the cruel, megalomaniac power of Qin Shi Huang,
unifier of China about twenty-two hundred years ago and its first emperor. Having ordered the warriors to be made to accompany his spirit
into the next world, he then killed most of the artists and had his scholars buried alive in a fruitless attempt to keep the place a permanent se-
cret. The scale of the display of a thousand warriors was impressive, yet
no two faces were the same as far as we could see.

In reaching Lhasa we achieved a near lifelong ambition, each of us
having read about Lhasa in Heinrich Harrer's *Seven Years in Tibet* (1953)
decades earlier. The bus took us through an extensive Chinese part of
Lhasa before we reached the old town and the Xiangbala hotel, just a
three-minute walk from the market around the corner. Twice we strolled
around the market, and we bargained for a couple of prayer wheels, a
singing bowl, and a horn. Ruddy-cheeked Tibetan women dressed in
cloaks, shawls, and head braids sat at their stalls laughing or smiling.
"Hello, just looking," was their soft entreaty.

We set our sights on the Potala Palace, Jokhang Temple, and Barkhor
market. The Potala is even more impressive on a visit than in the images
captured by photographs. Built on the rocky side of a mountain, the
building is full of chambers, rooms, and small temples, which in turn
are full of an extraordinary richness of paintings, Buddha figures, stupas,
bejeweled golden caskets, and so forth. The richness is overwhelming;
we moved too fast through it all with our guide, and the images started
to blur together. Which was the Dalai Lama who got lost? Which was
the one who . . . ? I liked the library, with its storage of history in row

upon row of scrolls (like chromosomes) in long boxes. The last (current) Dalai Lama's apartment was also impressive in its simplicity and tasteful design. We learned about the different Dalai Lamas, the blue sect and the yellow sect.

Our experience in the smaller and equally crowded Jokhang Temple in the afternoon was similar, with the difference that monks were chanting their verses from scrolls, and young men and women were performing a chanting dance on the roof that simulated the breaking of stones as effectively as if it were a choreographed ballet. Incongruously, a monk pulled out a cellular phone from beneath his robes. "Taking a call from heaven," as one member of our group put it. We learned that donations to the monks in both places end up in the hands of the Chinese government. One member of our group saw a Chinese soldier help himself to a few yuan notes.

In the evening we had a delicious meal at a Tibetan restaurant, although I must add that yak butter tea is decidedly not my, well, cup of tea. Singers and dancers in traditional clothes accompanied the meal, the men vigorous, expressive, and loud, with deep voices, the women decorative, not moving much, with high-pitched, almost adolescent voices. The evening ended with the audience being invited to join in a lusty song and dance on the stage, everyone holding hands in a line. We were two of the first to join in and finished completely out of breath. Rosemary loved the deep voice of her other neighbor.

On a visit to London at the end of 2002, we spent a fascinating couple of hours at an exhibition of Aztec art and history at the Royal Academy with Lukas Keller and his wife, Sarah Stead. Lukas had been a postdoctoral fellow with us in Princeton, and our friendship with Sarah extends back to the time we lived in Montreal. Much of the exhibition was new to us, because the material had been discovered after our Mexican experience in the 1960s. In fact, the exhibition was somewhat centered on the Templo Mayor, discovered in Mexico City in 1978 and excavated as far as the water table allowed. Some large artifacts were surprisingly well preserved—for example, a terra-cotta figure of a half-flayed human sacrifice. I do not remember learning before that the Aztecs valued feathers and flowers more than gold, which they considered to be the excrement of the gods—in other words, excrement of the highest quality!

This was a prelude to the Royal Society's 341st anniversary meeting in the afternoon and evening, at which Rosemary and I received Darwin Medals. No award could have given us greater pleasure—such is the magic of Darwin's name. The whole event—with reception and good conversation afterward—was thoroughly enjoyable. Bob May, presiding, paid us the compliment, "a long-happily married couple who work so closely together their individual contributions cannot be identified." That is a fair statement. We do work closely and know what each of us did, but most of the time we cannot identify the original source of an idea, and fortunately it does not matter. We think it is the first such joint award. The honor was particularly fulfilling for Rosemary, since she was not a fellow of the Royal Society at the time.

Standing in line waiting for the champagne, four medalists found themselves together and compared medals. This was a great opportunity for one-upmanship, because one medal was bronze and small, ours were silver and large, and the Croonian Medal was gold and even larger (and heavier). Its new owner, Sir Richard Peto, gave a faint, Olympian smile, without comment, when I told him what the Aztecs thought of gold.

In 2003 I was inducted as a newly minted fellow of the Royal Society of Canada. The meeting was in the National Gallery auditorium and the banquet in a splendid space under a glass dome at the Canadian Museum of Civilization. We stayed with Rita Palin—a friend from our years in Vancouver—and took her along to the ceremony as a guest. The setting of the evening banquet in the museum was stunning. A Haida Gwaii village front stood as the backdrop to the dais where speeches and awards were given. The First Nations section of the museum is outstanding, as is the architecture of the hall that contains it. For added atmosphere, a tape-recorded raven called out and commented on us mortals. The raven is a potent, mischievous figure in mythology in western Canada, a teller of tales and a keeper of secrets. Only the mist and rain were missing. Rosemary's induction as a foreign fellow followed in the next year.

We returned to Asia for two weeks, this time for our first trip to Japan. The visit was most rewarding thanks to the courtesy, attentiveness, and generosity of our hosts in Tsukuba (at the fiftieth annual meeting of the Ecological Society of Japan) and Kyoto (a symposium entitled "Toward the Integration of Biodiversity Studies" at the Center of Ecology, Kyoto University). The conference center and immediate vicinity in the Tsukuba area are modern, and for us Westerners it was a kind of stepping-stone on the way to Kyoto, which is old. Both visits were enjoyable, but of the two, the week in Kyoto was the more rewarding because we were able to strike a satisfying balance between the two-day meeting, visits to temples and shrines, and walks in mature forests. Unfortunately, the emergence of the famous cherry blossoms was delayed, and we missed it, even on our last day when walking along the famed Philosopher's Path by a small canal. We spent separate days on Mount Hiei, after visiting the eighth-century temple Enryaku-ji, and Mount Kurama, the home of many temples and shrines on the south-facing slope. Coming down from Mount Kurama, our two graduate student hosts chose a lovely streamside inn for lunch, and there, in the traditional surroundings, we had the best meal of our whole stay. I could not cross my legs, but this seemed to cause no social embarrassment. Japanese politeness is legendary.

Among the numerous temples of Kyoto, perhaps Ginkakuji (known as the Silver Temple) is the most impressive because the setting at the base of a mountain and the gardens are so beautiful. Our kind driver-host, Dr. Takakazu Yumoto, took us to Nara to see its deer park and temples. There the Shin-Yakushiji temple, founded in 747 and miraculously and uniquely intact despite the ever-present threat of fires, has as a centerpiece a supremely serene Buddha carved from a single block of *Torreya* wood, surrounded by twelve figures, mainly military ones.

In 2004 I gave two talks in honor of Ernst Mayr, icon of evolutionary biology, the first at Harvard to the Nuttall Ornithological Club and the second at the end of the year at the Berlin-Brandenburg Akademie der Wissenschaften in Berlin. He was very much present at the first: physically frail but mentally alert at ninety-nine and three-quarters, Ernst was in the front row of the audience. After the talk I answered questions, and then he was asked if he would like to say a few words. He rose to his feet gingerly, pulling himself upright, turned to face the audience, and with his back to me gave a perfectly composed, fluent, impromptu ten-minute speech on the history of his association with the Nuttall Club, replete with examples of new birds he saw, when and where. He and I had a short and typical conversation afterward, in which he insisted that evolution would take place only in small populations—I could hardly demur—and gave his blessing to the Daphne study. After photographs were taken, he was taken home to rest and was unfortunately too tired to join a group of us at supper. He lived to age one hundred.

Three days later my left hip was replaced with a titanium one by Dr. David Mattingly in the New England Baptist Hospital. I had been looking forward to this day for a couple of years. Back in Princeton, my recovery was helped by the rich stimulation of the seventeen-year cicadas and their domination of the sound waves. We had been fascinated by the previous emergence in 1987, at the end of our first year at Princeton and at about the time we bought our house. In 2004, emergence began on May 12 and continued for six weeks. We had fun conducting a study across the road from home, tape-recording and photographing them, seeing how far they would fly on release and how fast (not very), and marking and translocating them to see whether they would return to their point of capture (they did not). Much of this was done with an undergraduate, Uta Oberdörster, and then I continued with the fast-flying annual cicadas that emerged later. I enjoyed returning to my Cambridge roots by immersing myself in the physics and engineering of sound production and flight, profoundly different from studying ecology and the evolution of Darwin's finches.

18

New Discoveries on Daphne

AT THE BEGINNING of the century, we had started to entertain thoughts of finishing field research, for reasons of health not science, but one decade later we retired from teaching and were continuing on residual funds, knowing that field research would end soon. In fact, the 2012 season—our fortieth—was to be the last. By then we had learned so much more about the finches we had been studying on Daphne and had decided it was time for synthesis, to put all our discoveries into a book. We accomplished this two years later (Grant and Grant 2014).

At each stage of research, we had clear questions that we attempted to answer. Our main discoveries, however, were unplanned. As Peter Medawar, the distinguished Nobel Prize–winning immunologist, once wrote, discoveries are not premeditated, and he gave three examples from the biomedical literature to illustrate the point (Medawar 1984). We have often said that none of our discoveries could have been put into a research grant proposal as a goal for future research. For example, if we had proposed to document and interpret natural selection in the 1970s within the standard three-year period of a research grant, our proposal would have been laughingly dismissed. Ecological events are often unpredictable, especially in the Galápagos archipelago, with its erratically fluctuating annual weather patterns, and the only way to prepare for the unexpected is to think it might happen and to be opportunistic when it does. This requires being imaginative and attending to empirical detail. And that is how we were able to take advantage of the drought of 1977 and its effect upon the finches, as well as the two discoveries I will now describe.

In the first decade of the new millennium, we witnessed the origin of a new species! It happened like this. A large male finch immigrated to Daphne in 1981, and two years later it bred with a *Geospiza fortis* female. Trevor Price had banded it with the number 5110. We followed Trevor in thinking it was a particularly large *G. fortis* and supported the opinion with a numerical analysis of its microsatellite DNA markers. By following the fates of all banded birds, we gradually became aware that the immigrant bird's descendants were breeding among themselves. In other words, they were behaving as a separate species. The males even held territories next to each other (chapter 14), and they sang an unusual song, had glossy black plumage, and were distinguished by their unusually large size. We called it the "Big Bird" lineage (Figs. 18.1 and 18.2). The patriarch, 5110, lived for a remarkable fourteen years, which is not far short of the longest life of a finch in the whole study (seventeen years for a female *G. fortis*).

Rosemary and I had each written in our notebooks that 5110 reminded us of *G. conirostris*, the large Española Cactus Finch from the island of that name. The idea that this exceptional bird had flown halfway across the archipelago to Daphne from the southernmost island seemed too far-fetched to be realistic. However, we should have adopted the advice from Rosemary's father, who, when asked how he was so successful in diagnosing illnesses as a medical doctor, replied, "I pay attention to exceptions" (B. R. Grant MS)—indeed, 5110 was a *G. conirostris* and had flown from Española. As described in chapter 26, we found this out years later when our collaborators in Sweden—Leif Andersson and Sangeet Lamichhaney—sequenced the complete genomes of 5110 and several of its descendants and compared them with the genomes of all other species throughout the archipelago. If ever there was a eureka moment, a single moment when we said, "Ah, now I understand," it was when we heard the news of 5110's origin on Española. One of our regrets in ceasing fieldwork was not being able to follow the fates of these fascinating birds. Nevertheless, up to the sixth generation,

FIGURE 18.1. **Upper left**: A member of the Big Bird lineage of finches, generation 4, on Daphne Island, Galápagos. **Lower left**: *Geospiza fortis* (both from Grant and Grant 2014). **Right**: The first record of the immigrant Big Bird 5110, from Trevor Price's notebook, 1981.

FIGURE 18.2. A member of generation 3 of the Big Bird lineage of finches feeding a fledgling, on Daphne Island, 2006. (From Grant and Grant 2014, Fig. 13.5.)

they were still breeding among themselves and, in so doing, still behaving as a separate species. Opinions differ among scientists on whether it should be recognized as a new species after so few generations.

In all respects, this was an amazing discovery. Having begun our study investigating how two species are produced from one through a process of *fission*, we stumbled on the astonishing and completely unexpected process of two species giving rise to a third by *fusion*. It became an iconic exemplar of homoploid hybrid speciation, that is the formation of a new species by hybridization without a doubling of the number of chromosomes, as happens in many plants and a few animals. And what a piece of luck to have witnessed it. This point was driven home by a schoolgirl, at the end of a seminar we gave in Oregon, who asked: "If the original Big Bird colonist had been a female, there would have been no independent lineage, would there?" True, because the female would have bred with a *G. fortis* male, and then the hybrid daughters they produced would have bred with *G. fortis* males that sang a song like their father, and the *G. fortis*–singing hybrid sons would have bred with *G. fortis* females. Although obvious in hindsight, we had never thought of this. Until then, we had not appreciated the full extent of our good fortune. But we did know that if the male colonist had died before the age of seven years, we would never have had a sample of DNA to solve the riddle of its origin.

That the presence of one finch species can affect the evolution of another was our second major discovery, and it illustrates the important principle that evolution is contingent upon preceding conditions and not solely dependent on current ones. As with the Big Bird lineage, the origin of a significant event occurred more than two decades earlier. The Large Ground Finch (*G. magnirostris*) established a breeding population on Daphne in the remarkable El Niño year of 1982–83 (chapter 14), when two female and three male vagrants stayed to breed. Over the years, the numbers of Large Ground Finches built up, so that by 2003 there were as many as 350 individuals on the island. They remained apart from *G. fortis* and *G. scandens* and did not hybridize with them. When a few *G. fortis* and *G. scandens* males copied a *G. magnirostris* song, they were repeatedly harassed by the much larger *G. magnirostris* males.

Then came a very severe drought. Scarcely any rain fell in the years 2003 and 2004, and more than 90 percent of the abundant *G. fortis* population starved to death. It was a repetition of the severe drought in 1977, of heavy mortality and natural selection, only with a major difference: now small birds had the advantage over large birds, and the average beak size of the population declined precipitously. The reason for the high mortality was the large *G. fortis* individuals were outcompeted by the even larger *G. magnirostris* for the dwindling supply of large and hard seeds, principally those of *Tribulus*. The *G. magnirostris* population also suffered; its numbers declined, and by the time rains resumed in March 2005, only thirteen birds remained on the island. Their food supply was almost exhausted, and they were on the way to local extinction. Only the regeneration of the food supply after resumption of the rains halted the process and then reversed it.

The episode revealed how the environment changes, how those changes may be reversed, and hence that natural selection can oscillate in direction.* It provided us with a clear-cut example of the divergence of species when they come together and compete for food under stressful environmental conditions: character displacement (chapter 14)—not immediately, but as much as twenty-three years after *G. magnirostris* first bred on the island.

We were not able to be on the island in 2005—the crucial year to find out who survived and who had not—because I had colorectal cancer, and after an operation to remove it, in Corvallis, Oregon, I needed a three-month convalescence while undergoing a mild chemotherapy. Fortunately, our colleagues Ken Petren, Lukas Keller, and Uli Reyer substituted for us for two weeks, as did Thalia for a shorter period, and together they found every banded finch on the island by walking all over it every day. That is how we were able to reconstruct the effects of the prolonged drought. The only members of the Big Bird lineage to survive were a brother and a sister. Remarkably, they bred with each other without evident ill effects of close inbreeding, so did their offspring, and the Big Bird lineage prospered.

* See footnote on page 201.

The research gained a terrific impetus in a new direction from a collaboration with Cliff Tabin, Arhat Abzhanov, and Ricardo Mallarino at Harvard. They are geneticists who study development of vertebrate animals. Cliff telephoned me and proposed a collaboration to investigate how the different sizes and shapes of beaks are controlled during their development. We knew of his work, and admired it and wished we could do the same kind, so when the call came unheralded, it was easy to answer unhesitatingly with an enthusiastic and unqualified yes. Over several years, the trio found key elements in the control of beak development. Put simply, a molecule labeled BMP4, short for bone morphogenetic protein number 4, controls how deep (or high) and wide the beak will grow but not how long. BMP4 coordinates cell development in beak tissues and is referred to as a signaling molecule. Length is governed by biochemical interactions involving a different signaling molecule, calmodulin, at the same time but in a different place (the beak tip). Beak width, together with length but not depth, is modulated by another biochemical pathway a little later in a different place in developing bone. These findings help us understand how differences in beak shape can evolve through a change in one dimension independent of the others. They constituted a third major discovery, but unlike the two discoveries on Daphne, they emerged from highly focused, targeted research: a search for relevant genes.

Discoveries on Daphne were precariously dependent on our good health—both physical and financial—and on our strong motivation to understand the dynamics of actually evolving populations. As we entered the twenty-first century, we oscillated between "we can go on forever" and "no, we can't," and then back again to "yes, we can." The Indian author R. K. Narayan once wrote that by about the age of eighty, the voices of complaint from various parts of the body become increasingly insistent in calling for attention and impossible to ignore. The early years of the century were the problematic ones for us.

In the year 2000 Rosemary suffered from a swollen knee, and Daphne is not the place for that. I did most of the work, and then I strained my groin and neighboring muscles. In the next year I began to have problems with my right hip and could not help wondering whether this was going to be my last field season on Daphne with my original bones, or even the last season on the island ever—a depressing thought. We scaled back on banding finches (another mistake). The following year, despite creaks and groans in the joints, we began to think we could continue for many more years, although at reduced pace, strength, and vigor. In 2003 Rosemary arrived on Daphne with a creaky right knee, overcame that, and left with a creaky left knee. My left hip stayed supportive but unenthusiastic about the visit. By 2004 it was clear I would need a hip replacement, which I had soon after the end of the field season (chapter 17). I survived on Daphne with the aid of my German walking stick, trade name Leki; actually, it was made in the Czech Republic, hence a Czeki-Leki. It was a terrific help, as it enabled me to go on the steep slopes. There followed the cancer operation in 2005. We were not yet finished with ailments, for in the year of our return to Daphne (2006), I developed a puzzling strain in the groin, without warning or apparent cause, just before we left the island. It was an abdominal hernia, and yet another operation was needed to fix it. And there ended the era of creaks and groans, for in the remaining years of our study, we were fit, healthy, and, it must be admitted, a little more cautious when moving around on the island.

Craig Venter, famous for his work in sequencing the human genome, visited us on Daphne in 2004. Guided by our son-in-law Greg, he arrived on our second day there to include a visit to the "island of evolution" to help promote his global survey of microorganism diversity. He told us a fascinating story of discovering more than a million new genes in at least eighteen hundred new species of marine microbes off Bermuda. Despite his occasional good-natured boasting, we hit it off well and enjoyed the visit. Our enjoyment had nothing to do with the bottle of champagne he opened in his first ten minutes on the island nor with a fine bottle of French wine he left for us. He must have enjoyed the visit because he came back with a different entourage one week later, and again left us a bottle of wine. The best quote—relayed to us by Greg—was, "Greg, I wish I had in-laws as nice as yours!" "Beware snap judgment," I would have said. He offered to help us if ever we thought of sequencing the whole of a Darwin's finch genome; it was extremely generous and tempting, but we were unsure. Perhaps we should have done it, but I think we would have relinquished control of the direction of our research. Instead, we waited for ten years.

Just as evolution is unpredictable because environmental fluctuations cannot be predicted, so too is the social and political climate unpredictable in Galápagos and on the mainland. The Charles Darwin Research Station was under financial stress and closed its *comedor* (dining room) in 2003. This was a great pity, as it had been a place for excellent post-supper scientific conversations, going all the way back to 1973, our first year. A period of relative tranquility came to an end in 2004 when a strike by fishermen closed the research station. They were aggrieved at restrictive regulations governing their fishing that they believed originated with the scientists. They picketed the place and threatened to become violent but fortunately did not.

The year 2007 was our most difficult visit to Galápagos administratively and made us think of treating it as the last. The difficulties hinged on getting permission for our collaborator Arhat Abzhanov and his group even to visit the islands, let alone collect the embryos he needed for genetic work. Problems arose from a misunderstanding on the part of Raquel Molina, the relatively new director of the Galápagos National Park, who was not a biologist and, without reading our proposal, believed we were going to export live eggs to Princeton, hatch the finches, and release them! The research station people were unable to give us much help. Graham Watkins, the director of the research station, was preoccupied with the enormous problems created by an assertive fishing industry and lobby, uncontrolled tourism, and so forth. At the same time, the Ecuadorian Ministry of Environment was working out how to draft a new policy for permits to distinguish commercial from noncommercial scientific activities. The ministry suspended all exports of genetic resources, and we had to leave all our precious blood specimens behind when we left. Eventually they were sent to Yale by mistake and then forwarded to us.

Each year of fieldwork was different in some respect and had a character of its own. Even two drought years were not identical in their effects on the plants and animals. We labeled each year by something distinguishing it. The first never-to-be-repeated year (1978) was the year of the shark. According to the family, I was nearly eaten by one! I was treading water close to the landing, covered in shampoo and washing myself, when a reef shark suddenly emerged from the depths, opened its jaws as if to bite my head, and just as suddenly closed them and turned away. I did not see the shark but was very much aware of the family screaming at me to get out of the water. We joked that the Aquamarine shampoo must have contained a shark repellent. Months later I happened to read an article in the *Montreal Star* newspaper about how divers in the South Pacific defend themselves against sharks by squirting a substance at them from a rubber pouch attached to the upper arm. The article even identified the chemical ingredients. A quick visit to the shower sufficed to check, and sure enough the shark-repellent substance was in the shampoo: sodium lauryl sulfate.*

All other years were tame in comparison but interesting nonetheless. The year of the albatross (1996), named for a bird that stayed just a couple of yards away from the landing, was followed by the year of the fur seal (1998), so called because one arrived and climbed onto the landing platform. Both events were unique. The year 2001 was the year of the octopus. First, one nipped me as I sat in one of the pools in the Genovesa "estuary." Then on Daphne a larger one tried to seize Rosemary as she filled a bucket at the landing. She panicked and knocked it back into the sea with a couple of sharp blows with the bucket. "You missed getting a supper," I told her. "So did he," was her instant reply (notice it was *he*). On a third occasion, a large male sea lion emerged from the depths with an apparent mask or muzzle wrapped around its face. The sea lion kept trying to twist its head around to bite the pink-brown body of the octopus on top of its head, while the octopus app eared to be trying to shut the sea

---

* There are three versions, and they vary in the number of sharks (from one to six), their identity (Galápagos, White-tip, or Hammerhead) and their target (my foot or shoulder). All agree on the jaws opening and my apparent indifference.

lion's mouth with its ropelike tentacles. The sea lion won the battle in the end, but it took him at least five minutes to deliver the fatal bite.

If it had not been an octopus-rich year, it would have been the year of the biting caterpillars. Caterpillars were abundant on Genovesa in 2001, even more abundant than in the outbreak year of 1978, falling off the *Croton* and *Cordia* plants and crawling everywhere. These were apparently polymorphic—green and black or all black—though possibly two species. Only the all-black caterpillars bit us as they crawled up our legs. We learned they sequester a chemical from the leaves of *Croton* that gave our clothing a tie-dye stain. When they ran all over our clothes lying on the ground for a final drying, their excreta left little black stain marks over underpants, bras, shirts.

A unique feature of 2008 was the breeding of pelicans near the cliff facing Santa Cruz Island. Two large screaming chicks in one nest were close to two screaming chicks in another, while a single chick was in a third nest eighty feet away. We never saw the parents. The year was unusual in two other respects. At two o'clock one afternoon, a rare clap of thunder was synchronized with lightning over Daphne. This could have been the end of our project, as well as us. The second event, on March 7, was a terrific wind with near-horizontal rain one evening soon after we had entered the tent. I thought our new North Face tent was going to be ripped off the ground and blown away with us inside, but it withstood the onslaught magnificently. Two weather systems, one cool and the other hot, had met at the Galápagos, and their conjunction created atmospheric turmoil.

The weather in 2009 was quite different. The sea surface temperature started high and then fell, making the nights comfortable, even to the point of being cold; we had no blankets, we never do—it was March after all, so the nights should have been warm. The days were hot, and the combination of strong sun and cool sea gave rise to the mistiest experience we ever had anywhere in the islands apart from Punta Espinosa on Fernandina. When we got up one morning, all of Santa Cruz had disappeared; the fog crept eastward, and at one point even the base and top of neighboring Daphne Minor had disappeared. Needless to say, the farther islands of Santiago and Marchena remained out of sight all that day and on several other days. We saw Marchena only once. There were almost no caterpillars except, right at the beginning, a horde of them on *Sesuvium* plants that were everywhere for a few days and we dubbed "man-eaters" because they bit us. The island was visited briefly by a Peregrine Falcon and a Red-footed Booby, and, for the first time in several years, a *Myiarchus* flycatcher, which accompanied us on the plateau. Surprisingly, there were no owls, and we found very few pellets. All these unique or at most very rare events illustrate the variability of the Galápagos world.

We left early one morning on a flat sea in lovely light. At a distance I filmed Daphne full-frame, and then zoomed out to leave it as a small island perched on the horizon, like a memory, fading with the passage of time.

---

*Note from page 195: Differences between populations in space can be equated to differences arising in time. This was Darwin's important insight (chapter 10, 109) and is illustrated by *G. fortis*. Changes in beak size across the forty years on Daphne span a range of average values that encompasses differences among nine of the eleven populations of this species on other islands.

19

Retirement

TWO FORMER PhD students, Trevor Price and Dolph Schluter, retired us in 2005! Believing we were ready to retire, they organized a hugely enjoyable, two-day symposium and reunion ("Grantfest") at the University of British Columbia. Real retirement took place three years later—retirement from teaching and administration, not from research. The two reasons for retirement were my increasing hearing problems, which prevented me from giving lectures in the manner of Socrates interacting in dialogue with his students; and the needs of the department to have our salaries and hire new junior faculty. The imminence of retirement forced itself into our thinking in 2005 when my cancer was discovered, as we realized neither of us knew how much longer we would live, so perhaps it would be a good idea to enjoy the freedom that retirement brings. The future was uncertain.

After convalescing from the cancer operation, a normal year compressed itself into eight months, beginning with a very brief return to Zürich for an external committee meeting of the Institute of Zoology. Our summer was punctuated (or punctured) by the sad news that Jamie Smith had died of lung cancer. We were grateful for the good memories of Jamie at Oxford, later as a companion on Daphne in 1973, and on our last visit to Vancouver just before Christmas. He was a great friend and a model of stoical fortitude in the face of deteriorating health.

He would have appreciated the World Summit on Evolution sponsored by the University of San Francisco of Quito and held at the USFQ campus on the outskirts of Puerto Baquerizo Moreno on San Cristóbal in the Galápagos. It was supremely well organized by the rector Carlos Montúfar and Carlos Valle, who was once my student (chapter 16). USFQ flew in food from the mainland every day so that the restaurants in town would be sure to have enough for us every evening, and the university's catering department took up residence and gave us superb lunches. Most of the 150 to 200 attendees had never been to Galápagos before, and a trip to Galápagos was a gift from heaven. The mayor of Puerto Baquerizo surprised us when opening the meeting by making Rosemary and me honorary citizens of San Cristóbal (*huéspedes ilustres*), together with Charles Darwin (posthumously); I felt the great man was standing next to us. Rosemary and I had to do well in giving our talk on Darwin's finches, and fortunately, both of us were clear and animated. At the conclusion of the program, she and I were awarded honorary degrees from the university, as were Lynn Margulis, for her distinguished work on symbiosis, and Thomas Kunz, an expert bat biologist. We had to furtively put on purple gowns in a back room for the procession across the sand and onto the stage for the speeches and the awards, be-hooded, be-gowned, and be-sandaled.

From bad health to accolades, 2005 was a topsy-turvy year. In the midst of our activities at UBC, we received the news, out of the blue, that we had won the Balzan Prize for Population Biology. It was our first international prize. An especially pleasing feature of the award is that 50 percent of the money associated with the prize has to be devoted to the research, preferably in support of young people, four in our case. The splendid awards ceremony took place in the Swiss Parliament building in Bern. Rosemary and I each delivered a three-minute acceptance speech against a backdrop of a mural depicting local Alpine scenery. We each committed the mistake of sitting down in a chair in the top row that is reserved for visiting heads of state, such as George W. Bush. From the audience's position, our heads were apparently in the clouds! At one point in the social events that followed, we had a brief and uplifting conversation with the president of the European Convention on Human Rights. Only on such extraordinary occasions as this can scientists like us meet such extraordinary people undertaking such extraordinary social tasks. Conversations were regularly interrupted by unfamiliar people coming up to congratulate us. Foremost were members of the prize committee and women who approached Rosemary and thanked her most sincerely and movingly. Evidently, Rosemary stood as a representative, showing that women can achieve distinction and perform on the international stage just as well as men.

Such a message was needed at our next venue, a symposium to celebrate the 150th anniversary of the Park Grass experiment at Rothamsted in England. This unique ultra-long-running experiment was designed to address questions of nutrient enrichment on plant growth. Rosemary was invited to come but not as a speaker. In fact, there was not a single woman on the program. I mentioned this to someone, who replied, "There aren't any good ones." Seriously!

Back in Princeon I nearly missed giving a lecture to my undergraduate Evolution and Genetics class owing to nosebleeds, ghosts of my national service in Britain returned to haunt me. One laid me horizontal in the lecture room before the lecture. I was just able to organize my illustrations and explain to three students what might happen, when it did happen. The next thirty students just ignored me as they came in, as if walking over or around a professor on the floor was the most natural thing; the most laid-back professor perhaps! I caught snatches of conversations about next weekend, difficult boyfriends, football, parties, and so forth. The next year I gave my final semester of undergraduate lectures and ended with the theme of flexible behavior as a hallmark of human evolution and the need to use it to solve the problems we are creating in destroying the environment we live in, the future being more theirs than mine. And so ended my undergraduate lecturing career, on a high note of exhortation and as wound up as I have ever been. After Rosemary had ended her course on behavioral (evolutionary) ecology she felt a little sad, despite also ending on a high note. I scarcely felt anything. It was over, and I could barely absorb the fact. Retiring from teaching seemed artificial when we were so motivated and energized.

Returning from Galápagos in 2007, I narrowly escaped disaster. As I reached up to the top shelf of a bookcase in my departmental office, the whole of the metal frame and wooden shelf construction departed from its wall attachment and knocked me to the floor. My head narrowly missed the corner of my desk, and I fell clear of all the falling shelves and books except one: a biography of dead fellows of the Royal Society landed on my chest. Amazingly, I was uninjured. While others rushed around anxiously, I thought of headlines: "Crushed by the weight of learning," or "Professor Grant transferred to emeritus status by his own library." Funny, but also deadly serious; I have read of scholars who died in exactly that way.

We flew to Barbados to join Arhat Abzhanov, Jennifer Gee, and Céline Clabaut on a reconnaissance trip to assess the feasibility of a study of beak development of three finch (tanager) species related to Darwin's finches. We stayed at the Bellairs Research Institute of McGill University. Strangely, I had never visited the institute while I was on the McGill faculty. The visit was very enjoyable, almost a holiday, and successful in the goal of capturing finches. One morning, while I was standing at the mist net and taking out a bullfinch, our EEB colleague Simon Levin telephoned to say that I had been elected a foreign associate of the US National Academy of Sciences. I had been nominated several years before, and I had thought the active period must have expired. This, in combination with an intense focus on the trapping, probably explains why I remained in a numbed and detached state of disbelief for quite a while.

Then it was Rosemary's turn for honors, with election to the Royal Society of London. In the final morning before the meeting, I visited the exhibition *Impressionists by the Sea* at the Royal Academy of Arts in Burlington House, while Rosemary was instructed at the society on how to sign her name with a quill. When standing in front of a Claude Monet painting of sea crashing onto a beach, I made the surprising discovery that I could make the waves come alive by squinting at them through half-open eyes. Did Monet squint when he painted them? In the afternoon we were joined by five guests at the induction ceremony when each new fellow came forward to sign the fellows' book. The day concluded with tea, good conversation, and photographs on the balcony, followed by a splendid dinner at Veeraswamy with my sister Sarah and her husband, Alex; Rosemary's brother Andrew; and Peter and Pauline Checkley, friends from our days in Montreal.

In the lead-up to retiring we spent a furlough year in Vancouver. We drove there in our Volvo. Revisiting Ashnola in British Columbia—a favorite of ours in graduate days—was the peak experience of the whole journey. We missed the entrance to Cathedral Provincial Park and had to retrace our journey from Hedley to find it. The approach was completely new to us, as was the provincial park itself. We drove through familiar Ponderosa Pines along a gravel road of a First Nations reserve to the farthest campground and a little beyond, to the bridge over the Ashnola River, and there found the trapper's hut that we frequented on our climbs up to the Bighorn Sheep range some forty-six years before. This was a journey down memory lane, for sure.

On arriving at UBC, we quickly settled into Saint John's College. On the very day of our arrival, two young women approached us as we were returning to the college after a walk, and one blurted out: "We have just been talking about how we want to get married, and we saw you two, and we said how sexy they look. I bet you still enjoy sex!" "How did you know?" was my quick comeback. In Princeton we had a similar experience when a Hispanic woman, in her thirties, I would guess, stopped us and said, "I must tell you how much I enjoy seeing you two together! I love my husband, and I want us to live a long time and be like you!" "Why?" I asked another woman who had paid us a similar compliment. "Because you get on so well together, you look as if you have just met!" I am sure the real reason is Rosemary's three-hundred-megawatt smile, very different from La Gioconda's but just as bewitching.

We looked at Vancouver with a view to possibly retiring there but shrank back into our shells on realizing how expensive it would be to have anything like our Princeton house and surroundings. Visiting Vancouver was a fortunate move because we were on-site to help Thalia and family reestablish themselves after learning in Ecuador the distressing news that Olivia had developed type 1 diabetes. We found them a place to live, a doctor, and a dentist, and generally helped them get settled into an apartment and a school on campus.

In the department we occupied an office close to the one where we had met. The departmental secretary gave us the unexpected privilege of looking at our old files, Rosemary as lecturer and me as graduate student. A letter of recommendation from Michael Swan in Edinburgh for Rosemary repeatedly referred to her as an attractive young woman with strong academic potential, a socially acceptable type of comment in 1960 but not now—even if absolutely correct. We also went down to the shore, sat on a log, and had lunches there, looking out over the inlet and reminiscing over conversations in the past, about our strong attitudes toward educating children, for example, and those first tentative steps toward declaring ourselves in love.

The year ended with another burst of travel and lecturing. We gave an evening lecture at the impressive CosmoCaixa museum of science up on a hill above Barcelona, and enjoyed strolling in the city and dining on the sidewalks of Las Ramblas. This was a very pleasant return to European living, made even more pleasant by the generous hospitality of our hosts. Paintings in the Picasso Museum in a medieval castle-like building were educational. Almost without exception, the faces of his early portraits were at a slight angle to the artist, and the leading eye was painted much better than the other. This seems to be the origin of the style he developed later, of superimposing several views or planes of an object to depict multiple perspectives and achieve depth on a two-dimensional surface.

We should have gone to Washington, DC, in 2008 for my induction in the National Academy of Sciences, but instead flew to Zürich to receive honorary degrees, an embarrassment of riches. I was no longer *die Begleitung* (the attachment) of Frau Professor Dr. Rosemary Grant, as in 2002, but a fully independent scholar in my own right. We were the first duo to receive honorary degrees in the seventy-five years of the university's history. The memorable part of the ceremony was the piano playing of Mozart pieces by a young pianist, and the most memorable part of the social event afterward was an exchange I had with a chemist at lunch. He was the host of another honorary degree recipient, Roald Hoffmann, a Nobel Prize laureate no less. The host asked me, "How many honorary degrees have you received?" "Two," I replied. "My guest," he said, "has thirty-eight." The ratio of one to nineteen probably reflects our relative value to society, although mine might be somewhat inflated. I am sure my mother would have agreed!

Thalia and Greg were married outside the Asian Studies house on the UBC campus on her birthday. It was a simple ceremony under enormous Douglas Firs, damp and a little cold, while the sun did its ineffective best. Thalia was on edge. Greg was serene. Olivia read a lovely poem she had written. Her brother, Devon, wanted to go off and look for squirrels.

We greatly enjoyed the last few months in Vancouver, discussions with Dolph Schluter and colleagues, working on statistical problems in the department in the daytime, and going for dead-fish walks in the evening in the grounds behind the Museum of Anthropology on campus, or listening to talks or classical music from the Borealis String Quartet (they were extremely good, pre-professional).

On heading back east, we took the northern route through British Columbia in indifferent weather. The scenery along the Banff–Jasper Highway is majestic, the major reward from driving this route. Beyond British Columbia, the best parts of our trip were a visit on dirt roads to the Royal Tyrrell Museum in the Alberta badlands, and the rediscovery of the Grand Coulee campground east of Brandon in Manitoba. The Tyrrell, a magnificent museum of paleontology, is rich and informative, all the more impressive for being situated in the environment in which the dinosaurs roamed and their bones were found. We had missed seeing the Grand Coulee campground on our outward journey, and it had remained just a warm memory of a visit one afternoon in early May of 1973 after a long drive. Rosemary recognized the site as we drove past along the highway, and this was remarkable because it was not signposted and had been abandoned by the provincial government many years before. A man who had worked there for many years explained to us he had undertaken to restore the place that he loved. The site is of considerable significance as an annual meeting place of several groups of First Nations people and a place where American Bison used to be herded over a bluff and down into an enclosure.

It seemed that had we hardly got back to Princeton before we were off again, this time to Portland, Oregon, for Rosemary to give a plenary address to the American Ornithologists' Union (AOU). In June 2008 we had started the "Darwin season" (next chapter) in celebration of anniversaries in Darwin's life by jointly contributing to a small one-day symposium at the Royal Ontario Museum in Toronto. We were both lackluster, failed to scintillate, and left feeling we had let down our audience and hosts as well as ourselves. Rosemary was determined to do a far better job in Portland, and this she certainly did, to enormously long applause, a standing ovation from the back of the audience, and a remark from a questioner that it was the best plenary address he had heard in twenty-seven years of attending AOU meetings.

On September 1, we both retired. The department hosted a one-day symposium in our honor, and on the following two days, Rosemary and I hosted a two-day symposium funded by the Balzan Foundation, thereby completing our use of their funds. About three dozen speakers, family, and some friends (e.g., Kalffs from Montreal, Ulfstrands from Uppsala) stayed for all three days. As is so typical, Rosemary and I had to do a lot of running around to make this work, because the financial and logistical support from the chairman of the department, Dan Rubenstein, although indispensable, was not enough. The talks were given in a comfortable lecture room in the School of Public and International Affairs building. The mood was excellent, as were most of the talks (Brodie 2011). With the Balzan funds, we had invited four Ecuadorian students to participate with posters and short talks. One of them, Jaime Chaves, now has his own independent research program on Darwin's finches.

Our personal highlight was a daughterly speech from Nicola at the end of the dinner in Prospect House on the first day. It was dazzling; she told us afterward it was the first she had ever given. The theme was "My Family Are Not Normal," and the talk began with her saying she thought she was normal until she started to compare herself with other children and see, for example, their weird reactions to her garter snake, which escaped from a show-and-tell exercise at school. I was lightly roasted (i.e., ridiculed) in stories such as the one about her falling on the lava on Genovesa. From about a quarter mile ahead I had turned around and called out, "Are the binoculars all right?" Her speech ended with some very warm sentiments; it was a tour de force, eliciting some extremely complimentary and private comments from envious parents.

20

Celebrating Charles Darwin

THE YEAR AFTER we retired, 2009, was like no other and deserves a chapter to itself. Two anniversaries coincided, those of the birth of Charles Darwin, in 1809, and publication of *On the Origin of Species*, in 1859. They were celebrated all over the world with lectures, symposia, and academic fanfare. Darwin's finches ensured we were in the limelight. By the end of 2009 we had received ninety-five invitations to speak and had accepted and delivered forty-five of them, in an extraordinary variety of places and settings.

The celebrations actually began in June of the previous year, and by the end of 2008, Rosemary and I had given a total of twelve lectures between us, finishing in December with talks in Bogotá, Colombia, and at the British Trust of Ornithology (BTO) in Swanwick, Derbyshire. Of all these experiences, we probably enjoyed most our visit to Colombia. We stayed with Margarita Ramos (ex-Princeton) for only a few days but enough to do interesting things. These included visiting the magnificent Museum of Gold in the city and the Encenillo nature reserve, about thirty miles outside Bogotá, where Hendrik Hoeck, family friend and director of the Charles Darwin Research Station in 1978–81, had grown up. There we caught a glimpse of the remarkable Sword-billed Hummingbird, a bird with a beak longer than the rest of its body. The genes that control beak development seem to have got out of control!

The year 2009 began with a tour of Europe, first to Tenerife, which was new for Rosemary but for me was a return to the Canary Islands after an absence of thirty-eight years. We spent almost six days there, invited by the Fundación Canaria Orotava de Historia de la Ciencia. Rosemary gave a lecture at the University of La Laguna, and I gave another at the *fundación*. Our various hosts were extremely hospitable, courteous, and informative and took us to various places. These included Garachico, where a lava flow in 1705 had run down the hill, through and over the town, and into the sea; Icod de los Vinos and its remarkable dragon tree (*Dracaena draco*), believed to be about five hundred years old; and, on the last day, Pico de Teide. An endemic Tenerife Blue Chaffinch came out of the forest to make us welcome, while the scenery around the peak invited us to climb. A perfect beginning to a very busy year, it was a forerunner of things to come.

From Tenerife we flew to Milan via Frankfurt and switched from insular rural to continental urban. A joint Darwin Day lecture at the Museum of Natural History was our latest ever, from about 9:30 to 10:30 p.m. Before that, while everyone else enjoyed wine with their meal and protracted social interactions, we stayed sober. I had learned to avoid drinking alcohol before lecturing from an embarrassing experience as an undergraduate at Selwyn College. I was giving my first ever public talk, to the college Science Society on bird navigation with colleague Mike Young (chapter 5), and we each drank a glass of sherry beforehand to calm our nerves. My contribution to the talk was a supremely fluent five-minute sprint; the middle five minutes was a mile, and the last was a marathon and total mental exhaustion as I struggled to reach the finish line.

Our main reward for the lecture came the next day: tickets to a performance of Wagner's *Tristan und Isolde* at La Scala. The vice president of the Balzan Foundation, Carlo Fontana, was the director of La Scala for many years, and thanks to him we were seated right in the middle of the ninth row back from the orchestra. The singing, acting, and staging were excellent. The opera itself is not uplifting, even though it engaged our whole attention from 3:00 in the afternoon until 8:15 in the evening, as we followed its tangled philosophy of the futility of life.

On our last day in Milan, Suzanne Werder and Marcello Rossi of the Balzan Foundation generously gave us a guided tour of several churches and the cathedral. Leonardo's long wall painting of the *Last Supper* at the Cenacolo is a true marvel that represents the shock and consternation among the apostles when told a traitor was among them. We admired the mosaic of Sant'Ambrogio in the very old church (fourth century) dedicated to him and San Simpliciano, another old church of similar age. At this last one, Suzanne had arranged a meeting with the director of the associated theological college. Suzanne gave him a copy of our lecture, and in return he gave us a critical essay he had written on intelligent design and then a short, conducted tour of the church and the cloisters of an attached monastery. Discussing not-so-intelligent design was an unexpected and improbable ending to a day that began with a stroll around an exhibition of musical instruments in the castle built by the Sforza family and seeing Michelangelo's unfinished Pietà. The artist worked on it for thirty years, on and off. It was fascinating; the finished part was beautiful, while even the unfinished part expressed extraordinary compassion. No futility of life here!

From Milan we flew to London to attend a meeting of the Linnean Society. The next morning, we went to the Natural History Museum to see the exhibit on Darwin. The manuscripts and historical parts were displayed extremely well. I experienced a particular thrill on seeing the four notebooks and reading some key statements in the development of his ideas about the evolution of a new species, a voice from the past expressing an idea for the first time. However, in the biological part of the exhibition, our research on speciation in Darwin's finches was inexplicably reduced to a game of bagatelle! In the afternoon, together with ten others, Rosemary and I each received a Darwin-Wallace Medal at the Linnean Society. We were conscious of following in the footsteps of some great scientists, including Alfred Russel Wallace, J. B. S. Haldane, Hermann Joseph Muller, and Ronald Fisher. Mohamed Noor—the youngest among us—then gave a brilliant speech in reply, with thanks on behalf of all of us medalists, finishing with guesses about progress in the future, such as a return to the naturalist tradition of Darwin and Wallace after the enormous genome-sequencing activity has begun to subside. Well, maybe, if it ever does subside.

We returned to Canada, a one-day trip to Carleton University and the University of Ottawa. Our talk in the evening, to an audience of 350 to 400 mainly non-university people, was going well until the last picture but two, when the fire-alarm bell started to ring. Rosemary, unfazed, quickened her speech and reached a crescendo with the last picture, which gave a conservation message about keeping finches and their environments capable of natural change, printed atop a background of Volcán Sierra Negra, on Isabela Island, in flames and spewing forth molten lava. She finished like Boadicea or Joan of Arc, exhorting her supporters to follow her out of the hall—the only thing missing was a sword—and was greeted with terrific applause and vocal roars. We waited outside for security people to deal with the problem or pseudo-problem. I found myself talking to a man who asked whether I believed in God. Another told me he taught biology at Nipissing University, in North Bay, Ontario, and had driven four hundred miles to hear us; very touching and humbling, this sort of comment makes us always strive to do the best we can when speaking in public—the lesson I learned from Lawrence Bragg (chapter 3).

At the end of April, we went to Washington, DC, to be inducted as foreign associates of the National Academy of Sciences on the first day of the annual meeting. A big fuss was made of checking our identity to make sure we truly were who we said we were and not imposters off the street. *Right*, I thought, as I heard that; something clicked in my brain. After signing the book, I walked over to the genial president, Ralph Cicerone, as instructed, and after he had congratulated me ("Welcome aboard") and while our hands were still clasped in a warm handshake, I said quietly, "What would you say if I said I was an imposter off the street?" He threw his head back and laughed to the ceiling. This exchange was photographed, and to my astonishment the photograph showed the opposite: I was laughing to the ceiling, and Ralph was looking at me with deep suspicion!

The highlight of the meeting was an electrifying speech by Barack Obama. How extraordinary to listen to a politician, in this case a US president no less, speaking in plain but scholarly language in such straightforward, logical, and rational terms. His promise to boost spending on science and education despite current economic austerity was well argued and enthusiastically received. The speech ended as follows: "We are reminded that with each new discovery and the new power it brings comes new responsibility; that the fragility, the sheer specialness of life requires us to move past our differences and to address our common problems, to ensure and continue humanity's strivings for a better world." He then shook hands with everyone in the front row and left.

The next European trip encompassed joint talks at Stockholm, Lisbon, Barcelona, and Valencia, each event celebrating Darwin in a different style. The first was a two-day symposium at the Swedish Museum of Natural History, rounded off with very fine dinners and some entertaining singing at the Royal Swedish Academy of Sciences one evening and at the Vasa Museum on another. I shall now remember that the *Vasa* sank in Stockholm Harbor in 1628 because this is the date of our dining room chest, inherited from Rosemary's family.

In contrast to Stockholm, Lisbon was a completely new experience. We were in the hands of José Feijó, the organizer of a well-designed exhibition devoted to Darwin and evolution at the Calouste Gulbenkian Museum, the best exhibition of the year in our experience. We walked with him to the Castelo de São Jorge and gained a flavor of the city from many viewpoints overlooking pantile roofs and narrow lanes (*becos*). On the morning before our evening talk, we went to the Gulbenkian Institute for Science outside Lisbon. The occasion was a morning symposium of four talks on neurobiology given by members of an international committee. I felt sorry for the only one who did *not* have a Nobel Prize. Sydney Brenner was one of the speakers. I spoke to him briefly before the symposium and told him that as a student in my final undergraduate year, in 1960, I had listened to him deliver an inspiring set of four lectures on the whole of genetics. His instant reply was he would now give all of genetics in one lecture!

We continued to combine work with pleasure in Spain. The first engagement was a talk to the Institute of Catalan Studies in Barcelona. The building had been a convalescent home in the nineteenth century, and the lecture hall was once a men's dormitory (the first time in her life Rosemary had lectured in one!). We took a bus to Montserrat, which I had visited in 1957, but Rosemary had never seen. The weather was beautiful, and after lingering to enjoy a few minutes of singing by the boys' choir in the basilica, we seized the opportunity for long walks in the mountain above the monastery. We had been starved of exercise, and this fully compensated. The smooth columns and pillars of rocks dominate the skyline, just as the tall oak trees dominate the valleys. It was one of those delightful days full of pleasant surprises when nothing goes wrong.

We then traveled by train to Valencia to give the Peregrín Casanova memorial lecture at the university in late afternoon after a walk around the botanical garden and a filmed interview. Peregrín Casanova, a professor of anatomy at the university, had introduced Darwinism to Spain. I think this lecture was our joint best. Our reward was sparkling conversations with Juli Peretó, the rector, dean, and our host; and geneticists Andrés Moya and his wife, Amparo Latorre. We stayed in a restored university dormitory in the middle of the medieval city. Next day we walked around a market that was superb in variety and tempting produce from sea and land, animal, fruit, and vegetable, inviting comparison with the equally rich market in Turku, Finland (chapter 12).

As these excitements slowly faded, new ones replaced them, one after another, sometimes rapidly in this fast-moving year. Participating in a weeklong course on evolution for graduate students at Guarda in Switzerland organized by Dieter Ebert and Sebastian Bonhoeffer was one of the excitements. Twenty-six students from several European countries, plus three from North America, were divided into groups of five or six, and each group had the task of developing a research project/proposal, occasionally guided by five of us teachers: a total immersion in theory in a beautiful part of the world with one day off for a hike to enjoy it.

Visiting Oxford and Cambridge for two more talks was another excitement. At Oxford we were given the visiting seminar speaker hospitality for the first time: accommodation in Wadham College, dinner in the warden's (master's) house attached to the college, and so forth. We paid homage to the full-length portrait of Darwin as an old man, then gave our lecture. The visit was too brief, and so was the one-day visit to Cambridge. Apart from staying in a college (Selwyn), the experience was very different. I was there to take part in a Darwin Festival program organized by Pat Bateson. This was followed by a reception at Christ's College, where Darwin's room had been dressed up in the manner of his time by the historian John van Wyhe. Darwin himself, as a student, was present in the third court garden in the form of an arresting sculpture. We finished with a visit to the University Museum of Zoology, where the glass artist Tolly Nason showed us her replicas of the beaks of Darwin's finches, twenty times life-size and individually backlit by strong, ruby-red light to stunning effect.

Iceland was the next stage in the saga of 2009, unfolding with one day to recover from a short overnight flight, one day of two seminars in Einar Arnason's department at the University of Iceland in Reykjavík, one day with Einar driving into the hinterland, and a final day split between talks with graduates and an excursion to the lakes and waterfalls of Thingvallavatn, Geysir, and Gulfoss. Reykjavík had been noticeably transformed by the growth of spruce and birch trees in the gardens since we were last there forty years before. Einar drove us to Hekla and to Laki to see the amazing flood of lava that erupted in 1783 and spread for about twenty miles. The dust in the atmosphere that year reduced sunlight, causing crops to fail and the ensuing starvation and death of an estimated 25 percent of Icelanders. The eruption is believed to have cast a long shadow over continental Europe and contributed to the French Revolution through its effects on the crops. I spotted a Gyrfalcon high up on a cliff. Toward the end of the day, we drove into Vík í Mýrdal, at a southern point of the island, where we had been in 1968 to study puffins to complement the work of a graduate student (David Nettleship) on the same species in Newfoundland. We arrived in fading light but enough to see our old study sites, the black beach, and the Hotel Vík (renamed Hotel Puffin). What had been a small village of a dozen houses in 1968 was now a small town at least ten times as large and a popular tourist destination.

On our last day, in the company of Sigurdur Snorrason, a colleague of Einar and an expert on fish evolution, we were given fascinating introductions to the geology of various types of volcanic eruptions and to the enigmatic four morphs of char in Thingvallavatn—four steps on the way to new species. A farmer-fisherman and a friend, each dressed in not one but three pullovers, had put out nets before we arrived, and the two of them hauled in all four strikingly different types: large and small, sleek and deep-bodied, living in pelagic, benthic, or littoral habitats. Similar specialization in lake surface- or bottom-living is also seen in other fishes—sticklebacks and coregonid whitefishes as well as char— but four in a lake is unusual, at least in the temperate zone. After enjoying this illustrated lesson in diversifying evolution, an incipient adaptive radiation, we made brief stops at Gulfoss, the Niagara Falls of Iceland, and at Geysir to wonder at the chemistry of the brilliant cobalt-blue water.

From there we went to Oslo, where Rosemary gave the Kristine Bonnevie lecture, named for a remarkable and influential lady, a geneticist, and the first woman professor in Norway. She also happened to be born on the same day as Rosemary. Sadly, there was no time for visits anywhere after a symposium, and we left for a Darwin and island symposium in Minorca. We gave the opening address, and it went down well, and then I went down well with a cold that broke out under the whirring ceiling fan in the hotel bedroom. Several of the talks were on paleontology and very informative; for example, one was about an enormous fossil hare recently discovered on the island, and, for maximum contrast, another was on the remarkable *Myotragus*, a dwarf sheep fossil found on Majorca with a strange reptilian pattern of intermittent bone growth. On our last night some students entertained us all with a lovely dance and piano accompaniment in the city hall. In the last dance, two young women dressed as lizards did a mating dance, slithering sinuously over each other, extraordinarily lifelike, with a modicum of artistic license, and also funny.

From Spanish Minorca we continued on to Portuguese São Miguel in the Azores for a similar Darwin symposium, hosted by António de Frias Martins. The center points of our visit were excursions to two places, first to Furnas to see the hot springs and the second to the

Nordeste region to see the monomorphic bullfinch locally called the *priolo*. In 1972 we had stayed at the fine hotel in Furnas while studying chaffinches at a time when the Azores Bullfinch was believed to be extinct owing to destruction of its habitat. We passed the hotel and also Furnas Lake, where, on that first visit, we had been given chicken boiled in a pot placed in the hot mud. Pedro Rodrigues and Roberto Resende kindly took us to see the bullfinches in Nordeste. The native vegetation here is now recovering on the very steep slopes of the hills since the removal of introduced *Cryptomeria* trees. This is a striking testimony to what can be done in restoration, with dedication, patience, and determination. We encountered the bullfinches feeding close to the path and saw about twenty in the morning, a few adults with black caps and many more young birds without. Males are not red as they are in the bullfinch species of continental Europe but gray like the females. One thinks of redness as an essential part of being a male in the world of bullfinches, attracting females and repelling other males, but here the males manage without it. Whatever the reason, it is one of many examples of reduced brightness and conspicuousness in the plumage of birds on islands, which I first encountered in PhD studies on the Tres Marías islands (chapter 7).

Our last symposium of the year was at the ETH (Swiss Federal Institute of Technology) in Zürich, organized by Sebastian Bonhoeffer and Uli Reyer. It was superb, the best symposium of the year, partly because of the way it was organized and run and partly because all speakers were excellent and varied. I particularly enjoyed Craig Venter and Pim Stemmer on creating novel genomes and Bob May on a Darwinian analysis of why banks fail, but to single out individuals gives them undue prominence in a rich program that also included Andy Knoll (paleontologist), Svante Pääbo (geneticist), Linda Partridge (geneticist), and Tim Clutton-Brock (behaviorist). We gave our talk at nine a.m. on the second day, November 24, which I mention specifically because it was exactly 150 years to the day since the publication of Darwin's *Origin*. To help celebrate it Rosemary finished her part of the talk with the exciting Big Bird results, due to be published that day. In the evening we were entertained by a superb concert of Joseph Haydn (died 1809) and Felix Mendelssohn (born 1809), played by Sebastian (cello), his wife, Hannah (violin), and Oliver Schnyder (piano). Then we had a dinner at the Haus zum Rüden, a meal distinguished by the best roast duck of the century!

Our forty-fifth and last lecture of the Darwin year was given to a friendly audience at the University of Montana, the coda for the celebratory concert of a year. The day before, we visited the Bison Range reserve and other places with our hosts and past students Doug Emlen (Princeton) and Erick Greene (Montreal), and a group of students. The landscape was white with snow, the temperature at about 48°F, when the clouds cleared at noon, and all was peaceful, except for a bizarre interaction involving a Bald Eagle (predator), its prey (a heron), and three coyotes (scavengers). Darwin would have enjoyed it!

The year 2009 was our *annus mirabilis* for another reason. We were given the extraordinary honor of being chosen to receive the Kyoto Prize in Biological Sciences (Evolution, Behavior, Ecology, Environment) (Fig. 20.1). In Japan it is considered equivalent to the Nobel Prize for achievements in areas not recognized by the Nobel. And as with the Nobel Prize, there are many people equally deserving at the time the award is bestowed.

The prize is a product of philanthropy of the Inamori Foundation, led by Dr. Kazuo Inamori. Our experience began with an interview in Princeton with his brother, Mr. Toyomi Inamori, together with Mr. Kiyohiko Kigoshi as interpreter and Mr. Naoki Hirakawa as photographer. When Ranveig Jakobsen, our administrative assistant, said there were some visitors outside, Rosemary rushed out to bring them in and then, with them following, opened my door with a flourish—unfortunately, just as I was tying my tie. This was an ice-breaker; we all burst out laughing. They left after six hours, a record in their experience of interviewing candidates.

Their visit began with a forty-five-minute explanation of the procedure; then we were asked if we would come to Kyoto for the whole of one week of celebration and ceremony the next November, *if* we were confirmed as the laureates, and also whether we would attend a

FIGURE 20.1. **Upper:** Outside Eno Hall on the Princeton University campus on the day we learned of our Kyoto Prize, June 19, 2009. **Lower:** Recipients of the Kyoto Prize in Basic Sciences, awarded to us jointly by the Inamori Foundation in 2009.

symposium in San Diego for three days the following April; we said yes; and we were told, in that case, we were, tentatively, the 2009 Kyoto Prize laureates in Basic Sciences. Very different from a telephone call in the night from Stockholm!

We were politely quizzed on our attitudes toward such things as science and society, global deterioration of the environment, and so on. An expected question on spirituality, a main plank in Dr. Kazuo Inamori's philosophy, never materialized, perhaps because we had passed the spirituality test in our responses to those questions. We also passed a lightly disguised humility test, professing (correctly) modesty when it came to talking about our achievements. The most unreal part of the whole meeting happened when we told them that on June 19, the day of the official announcement of the names of the laureates, we would be traveling to Europe. They then filmed us accepting an award that had not been given to us, and would be given only if the board of directors accepted their recommendation. And then there was overt humor. The funniest moment came when we were asked, what was the most important event in our lives, and without a second's hesitation, I said meeting Rosemary; this elicited spontaneous laughter and murmurs of approval. Then Rosemary was asked about her most significant event, and because she said nothing for a few seconds, I said, "Obviously, it was not meeting me." This drew even louder laughs, almost guffaws.

Finally, the much-awaited Kyoto Prize week arrived, and it was truly a magical, fairy-tale experience of ceremonies and receptions, commemorative addresses, awards, a workshop, a banquet, a morning at the Saint Viator Rakusei High School for sixteen- to seventeen-year-olds, and an afternoon of "kids' science" at the Kyoto Municipal Science Center for Youth. Most of our immediate family could not come, but our son-in-law Ravi came alone, and later we were joined by Rosemary's brother John and his wife, Moira, as well as Jonathan and Debbie Weiner, so we were not without a support group.

The opening ceremony was as magnificent as it was moving. It began with a fanfare from a full orchestra designed to quicken the blood and heighten the excitement (Fig. 20.2). Speeches were given before and after

FIGURE 20.2. Kyoto Prize ceremony, Kyoto, Japan, 2009.

a Noh play. David Warren, the British ambassador, read out a message of congratulations from Prime Minister Gordon Brown. We gave our acceptance speeches and managed to bow to the presiding princess and the audience at the right time. It was very moving, and I surprised myself by almost choking up before the last sentence of my address. The ceremony ended with a chorus of twelve- to fourteen-year-olds, which brought tears to Rosemary's face, and an uplifting finale from the orchestra to send us forth. Before they left, two members of the girls' chorus—tall, smiling, and elegant in colorful kimonos—presented us with traditional embroidered Japanese threadballs signifying valued friendship. They are now treasured at home as lifelong mementos.

The banquet was splendid. About a thousand guests were present, in parallel rows, with the laureates at the high table, much as in an Oxbridge college. Sitting centrally between Dr. Kazuo Inamori and Princess Takamado, I did better than Rosemary for company. Communication with a smiling Dr. Inamori took place mainly but not entirely through an interpreter, who sat behind our shoulders, and whose English was amazingly good in structure and pronunciation for a person who had never been outside Japan. The Girton College–trained princess was highly articulate in idiomatic English, conversationally sophisticated, and loquacious. When the first dishes of French cuisine arrived, she warned me to eat quickly, because people would come up to say hello and interrupt the eating. She was interrupted much more than I was, handling encounters with great skill and wonderfully at ease. I did speak to the British and American ambassadors and several others who paid Rosemary and me generous compliments. There was some dancing and music afterward. We sat down at a table with the princess, and this gave Rosemary a chance to catch up with conversation. The princess told us she writes children's stories and is very keen on birds. She gave us some insights into what it was like to have married into the royal family; for example, she is permitted two weeks vacation per year out of the country, and only one of them is entirely private.

The kids' science meeting was a hilarious success, as was the meeting with high school students; ten minutes of us talking to the students, we were told, would be followed by ten minutes of questions. I began,

"Japan is a country with lots of islands. What is so special about—" and a hand shot up from the front row. I quickly realized we would have to reprogram our talk with a question-answer, question-answer format, and this was a terrific success, as we engaged the whole audience. Then came questions from the audience to us. "How old were you when you got married?" "Twenty-five." "Ooh, that's young!"

On the final day, Dr. Tohru Suzuki, our gracious and skillful guide throughout the week, also a plant scientist, took us to the gorgeous, otherwordly Koinzan Saihoji (also known as Koke-dera, or moss temple), a temple and garden restored in the fourteenth century. It was the most beautifully landscaped of all the temples we saw, and we took many photographs. We sat inside to listen to some chanting and wrote our wishes in ink on a wafer-thin wooden bookmark—completed on the back with our names and address. Then we went outside for a peaceful stroll with a few other visitors. The gardens are reputed to have 120 different kinds of mosses. They form an unmarked and uninterrupted green carpet beneath the trees, a forest fit for elves and gnomes before humans ever wandered here. Two men in soft boots swept the carpet with soft besoms. The counterpart on the other side of the globe is mossy Wistman's Wood (chapter 3), home to 120 species of lichens and spirits of the past.

After a splendid traditional lunch of many small and delicious items, including seaweed, sea cucumber, and ginkgo seeds, we visited the fifteenth-century Ryoan-ji, a temple with a tranquil stone garden for reflection, and seven rocks laid out in a pattern in a sea of raked gravel. We continued to two other sites. The first was Nijo-jo, a castle where the shoguns resided during the Edo period. The wooden walls of the various chambers of the castle were painted in the seventeenth century and are presumably touched up periodically, because the decorations are in good condition and strikingly realistic. For example, an eagle and a goshawk were depicted in picturesque pines, and when we walked outside, there, apparently, were the identical pines. A visit to the temple Kinkaku-ji was timed just right as the last rays of the sun lightly touched the golden pavilion. On the drive back through the town to the hotel, I recognized the stretch of river we walked along six years before.

Numerous Black Kites and crows were present now, as then, and so were the strikingly marked black and white Japanese Wagtails.

At the beginning of December, our EEB colleagues held a celebration in honor of our Kyoto Prize, with a reception at which two of Tolly Nason's glass beaks were unveiled to commemorate the event. The glass beaks, of *Certhidea* and *Geospiza magnirostris*, had been bought by the department, largely at fellow faculty member David Stern's initiative, and they were backlit with a ruby red light as in the University Museum of Zoology at Cambridge. I was a little overwhelmed, and I underperformed in responding to the welcome, congratulations, and unveiling. Room 10 in Guyot Hall was filled to overflowing for a special lecture that followed; I had never seen it so full. Again, this was thanks to David, who had the smart idea of inviting numerous high school students. They were so eager and enthusiastic, and some boldly contributed questions in the question period after the lecture. The lecture was on islands as evolutionary laboratories, superbly and amusingly given by Jonathan Losos, who has extensively studied the evolution of lizards on Caribbean islands. The day's celebration concluded with a dinner for the faculty at Prospect House, and warmed by wine, I was far more eloquent in describing my impressions of the Kyoto experience, from the emotions of the awards ceremony to the hilarity of the schoolchildren.

Kyoto celebrations spilled over into 2010, with another in San Diego in April. The celebrations comprised unexpectedly good press interviews at Point Loma Nazarene University; a gala night, for which I dressed in a tuxedo, where Rosemary and I spoke onstage for five minutes; a full-length lecture on biodiversity to an audience of one thousand at the University of California at San Diego; and dinners. Rosemary reached a high point after the lecture in response to a rambling question that finished with ". . . and Jesus lives with us today." She took the opportunity to launch into an exposition of the value of comparative religion and ethics in education as a means of learning tolerance as well as wisdom. Her reward was an enormous applause.

A couple of days after our return to Princeton, the department hosted the final Kyoto celebration: a two-hour symposium in McCosh 50 (a hall on campus) with two previous prizewinners, our colleague Simon Levin (theoretical ecologist) and Dan Janzen (evolutionary ecologist from the nearby University of Pennsylvania), Rosemary, and myself. This was followed by a reception and dinner in Princeton's Prospect House, and then a return to normal life.

Joel Achenbach wrote an entertaining essay about all this for the *Princeton Alumni Weekly*. "The Grants are almost comically warm and fuzzy," he wrote, "and still in great running condition, save a couple of dents in their fenders." "See yourself as others see you" is a good maxim, but I am still puzzling over the "comically fuzzy" bit. My beard, perhaps, for the fuzzy bit? For the comic part I turn to the more direct wisdom of a young grandson, Rajul: "Granddaddy knows when he is being funny but doesn't look funny, Grandmummy doesn't know when she is being funny but looks funny the whole time."

21

End of Field Research
on Daphne

RESEARCH ON DAPHNE became a long-term study of evolution incrementally, as one important result followed another. Other studies of bird populations begun at that time followed a similar course, without having an initial long-term goal. Most addressed ecological questions, such as what controls or regulates population numbers, and were not evolutionary. The situation has changed, and now an evolutionary focus is central to several impressively long-term studies of bird populations on islands around the world, from the warblers of the Seychelles (Pant et al. 2022) to the flycatchers of Sweden's Gotland to the song sparrows of Canada's Mandarte Island (Bonnet et al. 2022).

How do long-term studies such as these come to an end? Where continuity is key to success, there is no logical endpoint. There are always reasons to continue for one year or more, with the organisms dictating the schedule more than the scientists (Edmondson 1991). Nonetheless, they end, and often the reason is the money runs out. We would like to have handed the research to someone else but knew of no one who could continue when we took the decision to stop.

Our forty-year study of finches on Daphne came to an end in 2012. The lead-up to the final year was characterized by rain. On our last afternoon in 2010 on this arid little island, we sheltered from a downpour in the kitchen cave, but by seven thirty the rain had eased, and we made our way up to the tent. There we made the unpleasant discovery that water had made its way through the seams of the fly sheet and into the tent. We slept on more-or-less dry towels huddled together in the more-or-less dry middle third of the tent. The next morning, everything was not only sodden but filthy when we packed up to leave. Thankfully, there was no more rain. Lenin Cruz arrived on time with the *Pirata* to collect us, and we left at nine a.m. On the journey to Puerto Ayora, we were treated to spectacular views of at least a dozen cascades of water falling off the rocks between the northeast point of Santa Cruz and the Plaza Islands, not pouring over the surface but coming out of seams in the upper layers of the rock. This must have been what the pirates saw on one of their visits at the end of the seventeenth century (Dampier 1927).

The next year (2011) began differently and ended the same. There was predicted to be La Niña, and so there was, at the beginning. We called it the year of the mantas because so many of them were leaping out of the water, some with the maximum wingspan of about six or seven feet. Lenin confirmed they were everywhere and not just in the waters around Daphne. Did their abundance have something to do with the low temperature of the surface waters? Two Lava Gulls offered contention for the year's title. Their unexpected arrival one morning at breakfast time was a first occurrence. One gave us the evil eye as it walked up the slope to Rosemary and bit her toe, then immediately walked to the porridge pot and took one mouthful, as if by right, before I was able to chase it off.

When we left Daphne on the *Pirata*, we were 99 percent certain this was our last visit. It had been a dry year, and we had caught a lot of birds, enabling us to answer the question of whether *G. scandens* and *G. fortis* had continued to converge in body size and beak shape as a result of hybridizing: they had. By order of the national parks' office, we removed all the rotting bamboo canes, which had served as mist-net poles, and bits of table dating back to Peter Boag's years on the island (1976–78). We were sad to leave and did not look back. The end. No cause for celebration.

Our almost-confirmed plan to end the Galápagos research received a jolt at the end of February. On the twenty-fourth, Greg flew in to Baltra Island in such a huge rainstorm that it must have been archipelago-wide. He was able to make a quick trip to Daphne on the twenty-seventh, just to read the rain gauge: it registered ninety-two millimeters (more than three and half inches)! After long debate, we decided upon a return trip to coincide with the estimated peak of nestling banding in April.

We misjudged yet again! Unexpectedly, there was almost no breeding when we arrived, despite the accumulation of 150 millimeters (six inches) of rain in the gauge by then. Physiologically, the finches were as surprised by the rain as we were. Breeding began a few days after our arrival, and so we focused on the fate of the Big Bird lineage. There were more of these birds on the island than we had expected, at least thirty. *Geospiza fortis* had survived surprisingly poorly compared with *G. scandens* and *G. magnirostris*, and we wondered whether, perhaps, the Big Birds were being competitively squeezed, ecologically, by *G. fortis* and *G. magnirostris*. *Geospiza scandens* was about as numerous as *G. fortis*, and that had happened only once before. Another surprise was the 5–20 percent prevalence of avian pox on the faces of finches. Our most novel observation was a male *G. scandens* in fully black plumage singing a perfect Big Bird song at the top of the main path. It fooled us for a while. Finally, we netted it one still, windless morning, measured it, and took a blood sample. It was breeding with a "fortiform" *G. scandens* female, probably a hybrid. Not only that, by singing a Big Bird song, the *G. scandens* male was a potential mate for a Big Bird female, a source of extra genes and a possible pointer to future interbreeding.

The most amazing experience on this second visit was not the birds themselves but waterfalls, eight of them, cascading down the north slope of the crater on our third evening. Rain had fallen heavily on our first two evenings and was threatening to do so on our third; therefore, we ate early and at half past five set off up the main path toward the tent. We had almost reached it when a very strong storm burst upon us, so we tried to shelter under half-open umbrellas, overlooking the crater and close to the banding cave above the craterlet. From here we witnessed the extraordinary sight of the waterfalls, silvery white against the nearly black background. Six ran down in parallel and independently, while the nearest two fused at the bottom. The noise was deafening as torrents of water rushed over the ground and down the slope to the crater floor, while the wind howled, changing directions repeatedly, making it impossible to keep dry. Water fanned out in rivulets over the crater floor and then sank into the substrate, but not before leaving a pattern of little runnels on the surface. Features of erosion we had seen in daytime now made sense, yet another example of the importance of rare events. We didn't have a camera; it would have been too dark to use anyway.

In the last three years we enjoyed filming with our two Canon video cameras, mainly in afternoons when we were not netting, measuring, or following finches. I decided I had to use a tripod because my hand tremors were so strong, and although this was good for scenic shots, it was not good for moving targets like finches feeding on the ground. In the second field season of 2011, sphingid moths (hawkmoths) were abundant. Ever since boyhood I had been fascinated by hawkmoths (chapter 2) but had never seen so many, and here they were in daytime, flitting from one yellow starlike *Tribulus* flower to another, also visiting the yellow cup-shaped *Portulaca* flowers above and below the path from the crater to the landing. The most spectacular of the five species was a large green one with eyespots on its red hind wings. Best for filming was *Agrius cingulata*, a large, slow-flying moth, because it hovered at a flower for longer. Its tongue is more than four inches long and bent downward at about the midpoint. In high and erratic wind, the moth skillfully maintains station long enough to obtain nectar by dexterous adjustments to its hovering—impressive aerodynamics. These moths provided an interesting comparison with the finches, as they had taken different routes to the same end point of coexistence. Like finches, they are ecologically diverse, not as adults but as caterpillars that feed on different species of plants. Adults are diverse in color and pattern, unlike the finches, and again unlike the finches their diversity evolved on the continent.

The larger of the two video cameras broke when we tried to recharge the power unit from the solar panel. Then our one and only tape recorder broke, but in both cases most of the work with these two machines had been done. A few days before we left, we scored our best success with the other camera, a three-and-a-half-minute close-up sequence of a *G. magnirostris* in black plumage whacking a sphingid larva on a rock, then cutting it up into bite-size mouthfuls and eating them.

In the following few years, we had fun in the New Media Center lab at Princeton converting the footage into eight films for general education purposes (chapter 23): six on the finches, one on the moths, and one general film of life on Daphne.

The last field season (2012) was only two weeks long, like the first in 1973, but we managed to pack into it four intensive days of netting and plenty of walking and climbing in search of new Big Birds and banded birds in general. We caught two Big Birds banded as nestlings in the previous year and two new ones and observed several more, and found that recruitment was low but survival was high, as in the previous year. This was the year of abundant colonial tunicates, or salps, which we had never seen before, in the sea both off Daphne and elsewhere, as Greg informed us. What was unique about the fortieth field season? Finches perched on us more frequently than in any other year, with no sign of greater wariness as a result of repeated exposure to us. In the early years, *G. fortis* individuals were the only birds to enter our kitchen cave while we were there, but now as many young *G. scandens* as *G. fortis* shared the cave with us. A particularly bold and exploratory *G. scandens* gave me an excruciating experience by biting a nipple! By doing so, it established a peculiar symmetry for us visitors, a way of reminding us of the first year, when a barnacle nipped Ian Abbott's testicles at the landing. These acts of defiance from the owners of the island remind us of our place on it as visitors.

We left Daphne on a beautifully clear morning and took photographs as it gradually receded from view. We finished in Puerto Ayora, as we did on several previous occasions, staying with Thalia and Greg in their house on the other side of the estuary and eating fresh tuna from the market, sold at two dollars a pound, helped down by red wine.

"In our opinion," wrote George Oster and E. O. Wilson (1979, 314), two prominent evolutionary biologists, "the way forward in evolutionary theory is not through the formulation of global statements about the evolutionary process, but through the prudent choice of paradigmatic examples that permit the role of natural selection to be analyzed with unusual clarity."

We chose that route. Darwin's finches are a paradigmatic example of a young adaptive radiation of ecologically diverse species from a common ancestor. Speciation and ecological adaptation occurred repeatedly over one million years, resulting in the gradual buildup of the diversity of species we see today—and perhaps more if some species have become extinct. Daphne's finches are a paradigmatic example of the main driving force of the diversity—natural selection: one paradigm nested within another.

Although it is only a single island, Daphne is a microcosm of the whole radiation. Key events of species formation and coexistence occur when populations come together on an island like Daphne. The finches there have revealed the processes that we think have probably occurred numerous times in the archipelago, beginning with evolutionary divergence of populations of the same species on different islands and culminating in coexistence of species on the same island with little or no interbreeding. Events on Daphne have shown us that to coexist, finch species must differ in feeding ecology and choice of mates, with no interbreeding (*G. fortis* and *G. magnirostris*) or with only a little (*G. fortis* and *G. scandens*). Even when there is enough food for both, two species will not coexist if they are strongly similar in appearance and song (*G. fortis* and *G. fuliginosa*), for then they will interbreed freely, and the rarer one will become absorbed into the other population. On the other hand, interbreeding of two species can produce a third. Whether or not this has happened repeatedly in finch history or just once (on Daphne) is in the realm of conjecture. From bathymetric maps we can say, however, that Daphne-like islands formed many times over the past million years with the rise and fall of mean sea level, and it is scarcely credible that we have witnessed the one and only initiation of hybrid speciation.

Continuous long-term studies of nature in one place are valuable because they yield discoveries and insights that are impossible or unlikely to be obtained in the short term. For example, we needed more than a couple of decades to establish that finches can live for as long as seventeen years. This was important information when we sought to understand why finches vary so much in fitness, as measured by their success in procreation over their lifetimes. We gained important insights about evolution from events occurring five, ten, eighteen, and thirty-two years after we began.

In a well-chosen system with important questions to address, the value of repeated observations is multiplicative and not additive; each new finding can be interpreted in an increasingly richer context. That is why field biologists lament they did not start a decade or more earlier. Our chief regret in this regard was not sampling DNA at the beginning of the study. With hindsight we know we could have started in the 1970s. Luck comes to the prepared mind, it is said, and it also comes to those who persist and are in the right place at the right time in field studies such as ours. We have had that luck, several times over. Under the constraints of working in a national park, with little scope for controlled experimentation, we relied on observation and measurement. Our guiding philosophy has been to seek simplicity but to distrust it; to be creatively imaginative yet skeptical; to ask, how would we know we are wrong if indeed we are wrong; to seek a solution to a problem from several angles and not just one; to collaborate with others when in need of expert help; and to persist.

Looking back over the forty years on Daphne, what are the main take-home messages from the principal discoveries of the field research? Seven stand out: seven pillars of enlightenment. First, measurable evolution can occur as a consequence of natural selection in contemporary time in small populations of vertebrate animals, and is not simply an immeasurably slow historical process, as Darwin believed, or confined to agricultural pests or to bacteria and viruses in hospitals; and it may occur in fits and starts, episodic rather than continuous, fluctuating rather than unidirectional. Second, an exchange of genes between species through hybridization enhances the potential for evolutionary change, and even speciation, and is likely to have contributed significantly to the radiation of Darwin's finches. Third, a choice of mates is based on learning (imprinting) parental features and not, as far as we know, on genetically variable mate preferences. Such choice sets up and maintains a barrier to interbreeding between species, and furthermore, even when species in young radiations do interbreed the offspring are fit and show no sign of genetic incompatibilities of their parents' genomes. Fourth, environmental conditions fluctuate, and long-term studies of organisms are essential if we are to understand how those fluctuations affect them. Fifth, evolution is to some degree unpredictable because environmental fluctuations are unpredictable. Sixth, extreme events perturbing the environment, though rare, may powerfully affect the ecology and evolution of plant and animal populations. Seventh, speciation can occur when improbable and unpredictable events occur together.

David Wake summed up our forty-year project: "This research has resulted in an impressive body of work, which has had immense and, I think it safe to say, lasting impact on evolutionary biology" (Wake 2010, 361). Ryan Calsbeek wrote of our undertaking, "a team of field biologists [who] would spend decades of their lives camping on rocks in the middle of nowhere . . . spawned the single greatest legacy of modern field studies in evolution. . . . One may safely assert that no single research program has done more to illuminate the path of evolutionary field biology" (Calsbeek 2011, 828).

Working in Galápagos is a privilege, and we are enormously thankful that we had so much freedom at the beginning of our research, freedom to take the children into the field, for example. We were born at the right time. The freedom became increasingly constrained, as new regulations were necessary for the management of greater numbers of visiting scientists, residents, and tourists, while illegal fishing posed new and expanding threats to conservation. With freedom comes an obligation to respect the environment, and we responded by creating our own pre-camping quarantine system of washing, cleaning, and checking clothes, equipment, and food before arriving at an island to ensure we did not introduce animals or plants. The National Parks authorities instituted a formal and universal system of quarantine many years later.

A minor watershed event occurred at the beginning of the century, when a foolishly irresponsible group of visiting scientists misbehaved on Española Island, taunting sea lions and partying. They were indiscreet on returning to Puerto Ayora, were observed relaying their experiences to friends by e-mail in an internet café, and from then on, all visiting scientists were forbidden to take alcohol into the field, even beer.

After our final field season, I gave a one-day course on field methods of studying birds at the Galápagos National Park headquarters for rangers and for Charles Darwin Research Station personnel. With Rosemary assisting, I presented the course in my limited Spanish to about fifteen people. Afterward, I sent a set of digital images of all of Darwin's finch species to the national park for use by their guides.

On returning to Princeton in April, we learned that our research grant was completely spent. Having no money for research was something we had not experienced before. It was not the reason for our decision to make 2012 the last field season, nor was the increasing bureaucracy associated with permissions and permits. Both contributed to the feeling that we had chosen the right time to stop for much better reasons: a desire and psychological readiness to write the synthesis. Added to that were questions hovering in our minds about physical strength and stamina. I was facing the prospect of another hip replacement in the forthcoming year.

Before we returned to Princeton, we set off from Quito in pouring rain for the University of San Francisco's (USFQ) biodiversity field station on the Tiputini River. The first leg of the journey, a flight to Coca, was delayed, and so was the next, a canoe trip down the Napo River. There followed a drive across an area (block 16) controlled by the oil company Repsol, which took place without a hitch; and the final stage of our journey, a two-hour trip down the Tiputini, was wonderful. And "wonderful"—full of wonder—is the appropriate adjective for the next three days, spent in the company of our guide, Rameiro, in the daytime and about ten graduates and John Blake, a visiting ornithologist, at meals. We were in wet tropical rain forest, and what a contrast with the simplicity of Galápagos, thrilling and challenging in its complexity. An overarching question hovered in our minds and occasionally emerged in stimulating conversation: How much of the complexity can be captured by a distillation of the essential processes of evolution from a study in Galápagos, and how much can not? Only long-term study can answer the question, and it is needed, because there are signs of evolutionary change in Amazonian rain-forest birds in response to a warming climate (Jirinec et al. 2021).

One morning, at six thirty, we climbed a 130-foot tower and watched birds in the forest canopy for three and a half hours. On another, a canoe took us downriver, stopping twice to allow us to climb ashore in the Yasuní Nature Reserve, first to try for a close look at a two-toed sloth (we failed) and second to find a pygmy marmoset (we succeded). We hiked trails by day, always with our guide, and once by night to look at insects. And we took a short trip around a lagoon to see that weird and ancient bird the Hoatzin (Reilly 2018). The closest we came to danger, as far as we know, was about fifteen feet from the nostrils and eyes of a twelve-foot Black Caiman in the lagoon. In the river, two dolphins briefly broke surface in front of the canoe—more of a pale gray color than the expected pink—and then returned to the mysterious world of muddy brown water they shared with who knows how many species of fish.

Five species of monkeys allowed us to share their forest, some grudgingly: spider, howler, titi, capuchin, and squirrel monkeys. Among the more than one hundred species of birds we saw, the standouts were piping guans, Salvin's Currasow, manakins, many colorful tanager species in the canopy, toucans flying like missiles with wings, and a huge Crested Eagle, which we eyed at a great distance from the walkway in the canopy. Disappointingly, Anacondas and Jaguars were elsewhere.

The most common cause of death in the forest was not Jaguars or bushmasters, we were told, but falling trees or branches. This was easy to believe after a tree fell down with a terrifically loud crash in front of us on our porch while we sheltered from the rain, and another severely dented the corrugated iron roof over the dining area. We were in awe of our extraordinarily knowledgeable guide, Rameiro, and in awe of the bewildering diversity of the plants in the understory and the subtle changes in composition from place to place. We left with a strong feeling of ignorance of this magnificent, untouched biological diversity and deeply grateful for our brief but intense experience of it. The experience inspired some tongue-in-cheek verse (below).

Only when we opened our bags of clothes upon our return to Quito did we fully appreciate that we had been living in a saturated environment for the best part of a week. Eighteen months later, President Rafael Correa approved drilling for oil in the Yasuní National Park, in the Ishpingo-Tambococha-Tiputini oil field in block 31. Ouch! BBC News reported that 76 percent of Ecuadorians were against drilling. It was ugly politics and a huge retreat from a hitherto enlightened program to conserve tropical rain forests and their inhabitants.

A Walk in the Woods with Rameiro

Rameiro, what was that bird we just saw?
Was it woodcreeper, toucan, or swift?
It sounded like a cross between a hawk and a guan
if you are able to follow my drift.

A Plumbeous Pigeon, is that what you said?
Are you sure it wasn't a tinamou?
As it dashed through the air, I could almost swear,
it was the color of tiramisu.

The pygmy manakin, the one over there,
the green bird against a green leaf.
It does not move, it rarely flies,
its camouflage beggars belief!

The song it sings is a monotonous noise,
and therefore, it ought to be found,
but to make matters worse, and birders' curse,
it's a ventriloquial sound.

We don't tally numbers, we're here for the fun,
we have never before made a list.
But the total we know would be one hundred and four,
if ten of them hadn't been missed.

We have puffbirds and nunbirds, jacamars and antbirds,
manakins, euphonias, and wrens.
We would have seen more, a much higher score,
if a twig had not flicked out my lens.

We are up in the canopy, high in the trees
and can see they are not all the same.
So this is where all the birds are found.
Compared with below they are tame!

Macaws are huge parrots and colored like carrots,
their language is surely profane.
There's a Swallow-tailed Kite, and a fantastic sight,
an eagle the size of a plane!

My favorite's the Paradise Tanager,
Bright blue and green and red,
surprisingly bold, for one colored gold,
multicolored all over its head.

Rameiro, what was the bird we just heard?
Was it barbet, cotingid, or jay?
It had a green tail, it looked like a male,
its "song" was a donkey's bray!

And how do you tell all those antbirds apart,
when their plumages look all the same?
And how do you remember all of their songs,
when they too appear all the same?

The Green Oropendola, the Red-capped Cardinal,
and the Yellow-rumped Cacique.
Now these are a blessing to the naturalist who's new
to this tropical forest mystique.

Rameiro, you identify everything.
But I wonder if even you know,
why there are so many species here.
Is it because there's no snow?

And why are they so brightly colored,
except for the ones that are not?
Does it have anything to do with predation,
or do females like males that are hot?

Or perhaps there are so many species,
because forests have been here forever.
Tell us Rameiro, our naturalist hero,
you must know, as you are so clever.

We hoped we would see a Jaguar,
Fer-de-lance, and a coiled Anaconda.
But a dangerous snake, its absence doth make,
more enjoyable our tropical wander.

I'm here watching birds, and thinking of words,
so here is a little digression.
Almost as strange as the birds themselves
are the names they've been given in Latin.

For example, we know that *Ara macao*
is the name for the Scarlet Macaw.
Pionites melanocephalus and *Amazona ochrocephala*
are two other parrots we saw.

Spix's Guan, a piping guan,
and a Salvin's Currasow.
These are but few of the names that were new.
We knew much less than now.

Who was Salvin, and who was Spix,
and what were their particular tricks
to gain their fame by giving their name
to a bunch of birds in the sticks?

To judge from the names, Stolzmann was here,
and so was a guy called Schrank,

Lafresnaye, Richardson, and a lot more besides.
For them we have much to thank.

Thank you, Rameiro, for sharing so much,
for showing us monkeys and sloths,
explaining the value of medicinal plants,
and pointing out crickets and moths.

Thank you for showing us Hoatzins,
and others with very strange names,
caimans and falcons and two river dolphins,
that I'm sure were just playing games.

The experience was once in a lifetime.
Long after the names have gone,
I'll remember your guidance, patient compliance,
and your smile from the great Amazon.

22

An Epicurean Life

NICOLA GAVE ME a book, *The Swerve*, by Stephen Greenblatt (2011). It tells the story of how a book hunter, Poggio Bracciolini, found a long-lost manuscript of a poem by Lucretius in a monastery in Germany in 1417. *De rerum natura* (*On the Nature of Things*) was written in elegant Latin in about 50 BCE. I knew of Lucretius's proto-evolutionary ideas of natural selection from a quote in Haldane (1932) (Grant 2000), but didn't know that Epicurus—his philosophical hero—had an essentially modern vision of the world: trust in empiricism; be skeptical of the opinions of others; enjoy life, as there is no afterlife; have fun; help others whenever possible; do no harm unless unavoidable; and use the body and mind to the maximum extent for a fulfilling life. I worked out my own philosophy—a personal credo—along almost identical lines in my early twenties. We would have got on well together.

Our way of exercising body and mind and having fun in retirement took the form of giving lectures whenever invited to universities (Fig. 22.1), schools, conferences, and meetings of scientific societies, and then exploring the locations by hiking and indulging in our love of museums, art galleries, and archaeological sites. Stepping back into the past helps us to understand the present. As evolutionary biologists, this is what Rosemary and I and numerous others do when we reconstruct the history of our study organisms with their genomes. Culturally we do the same by visiting interesting places that reflect the past and asking not only how societies changed but why. A way of looking at the world outside science is conditioned by how our science has informed us and vice versa.

FIGURE 22.1. Giving a lecture at the University of Zürich, June 2012.

A career of sabbaticals strung together between periods of teaching on campus metamorphosed in retirement into a new career of permanent sabbaticals. In the five years after the *annus mirabilis*, our seminar visits took us on a random walk to Switzerland, Germany, France, Italy, Spain, Portugal, Ecuador, and Hawaii, and gave us opportunities to do much more than talk about Darwin's finches. In this chapter, I describe our experiences in Italy, France, Spain, and Portugal and conclude with the very different experience of hip replacements.

In the second week of May 2010, we traveled to Venice and then to Rome. As with so many trips, on the eve of departure we did not want to go, but as soon as we arrived, we were convinced we had done the right thing. Venice is magical, never more so than when it is approached by water taxi from the airport, looking like a mirage. We were the guests of the Istituto Veneto di Scienze and gave a lecture in a very old and well-preserved paneled room to an audience of about forty. We stayed in a comfortable guest room in the same building, letting ourselves out by the side door to go foraging. Next door, across a small street, was the church of San Vidal, where we heard a superb concert of Vivaldi music played by a vigorous group, the Interpreti Veneziani. The cellist Davide Amadio was outstanding, as was Andrea Bressan, a player of the *fagotto*, a bassoon-like instrument. The music propelled us to the museum of musical instruments (and the exhibition *Antonio Vivaldi e il suo tempo*) in the church of San Maurizio to examine the instruments more closely. One morning just disappeared as we walked around the painting treasures of the Gallerie dell'Accademia, until Rosemary had had enough of violent scenes painted on the canvases of religion. Two years later we returned to enjoy Venice all over again and under similar circumstances.

From Venice to Rome: The work we did to earn this experience was a lecture at the Accademia Nazionale dei Lincei (Palazzo Corsini) arranged by the Balzan Foundation. The setting was similar to that of the Istituto Veneto lecture, and the audience was about the same size, but I doubt if there was more than a handful of biologists among them. This very pleasant occasion was preceded by an interesting tour of the Villa Farnesina, built in the early sixteenth century for Agostino Chigi, a super-rich banker, who had the walls and ceilings decorated by expensive painters. The Galatea fresco by Raphael is the most prominent of a large and impressive collection. I cannot help noticing plants and animals in old paintings, to identify as well as to marvel at, if depicted well, as indeed the European birds depicted in these paintings are. The banker reputedly impressed important guests—including the pope—by ordering the dinner plates made of gold to be thrown into the Tiber River, at the bottom of the garden, after the meal, having previously ensured there was a submerged net in place to capture and recover them after the guests had gone.

On our one full day of exploration, we visited the National Etruscan Museum, the Forum, and the Palatine Hill, with much walking in between. The Etruscan Museum was extraordinarily rich, with much to learn and much to admire. Especially impressive were the early (seventh- to fifth-century BCE) pottery decorations in what I would have said is typical

Greek style but was subtly different, influenced partly by the Corinthians. Statues were rare, but one was magnificent, an incomplete figure of a striding man with vigorously contoured, taut-muscled legs.

Thanks to Giorgio Bernardi, patron of our earlier visits to Venice, we joined a workshop on molecular biology, genomics, and evolution at the Stazione Zoologica Anton Dohrn in Naples. Anton Dohrn had a unique vision to set up the station for research and to fund it by charging an entrance fee for people to see fish and invertebrates from the deep sea in large holding tanks. The workshop was rewarding, although we felt a little bit like fish out of water and partly out of our depth (excuse the mixed metaphors), as we were the only field biologists, and we were not ready to talk about the collaborative genomics work we had done with Leif Andersson (chapter 26). Our visit was a kind of pilgrimage to a place I had read about because when I was a student, I learned that J. Z. Young and other neurophysiologists went to the station every summer to work on the giant axon of squid, and so did Carl Pantin, my Cambridge zoology teacher. Later Evelyn Hutchinson spent a year doing postdoctoral work there, as described extensively in his memoir (Hutchinson 1979). I could appreciate the attraction of combining research with leisure.

To visualize what life was like in nearby Pompeii before it was buried in the ashes from the Vesuvius eruption in the year 79 CE, we immersed ourselves in the artwork (frescoes, bronzes, etc.) in the National Archaeological Museum of Naples before visiting the site. Pompeii is a very large restored area, and with little imagination we were among children playing in the street long ago, neighbors were gossiping, dogs were quarreling, people were lining up at an inn to be fed, and the noise and smells were palpable. The experience was vivid, but the next day we had the more powerful experience of visiting Herculaneum, where even the original beams and roofs were still in place on some houses. These archaeological sites are to the cultural anthropologist what fossils are to a paleoecologist or evolutionary biologist.

On the last day we visited Paestum. We had been advised that this ancient Greek site of the sixth and fifth centuries BCE with amazingly well-preserved temples would be the highlight of our archaeological experience, and so it was. Three temples of a Cotswold-colored

limestone dominated the site, only 20 percent of which has been excavated. Apart from the temples, buildings were little more than outlined
by low reconstructions of their walls, but it was enough to give an impression of scale and scope and appreciate how the Greek town was later
Romanized. Here, as at the other sites, there were almost no birds—
strange. In the museum, a famous fresco is adorned with a beautiful
depiction of a diver and a "symposium" of reclining figures flirting, drinking, and eating. The elegant diver, we learned, is interpreted as diving
into the unknown underworld, and the symposium figures are getting
themselves into that state by indulging in sensual pleasures.

A similar step back in time took us to Roman-occupied southern France,
Arles to be precise. The amphitheater, built by the Romans in the fifty
years preceding the birth of Christ, mainly at the time of Augustus, is
now in a good enough state of restoration to powerfully evoke the past.
When we were walking around the site and then sitting on the steps
looking at a modern replacement of the stage (or proscenium) and the
two columns standing behind it, we could easily imagine a play performed before hundreds of spectators. However, we were completely
unprepared for what we subsequently learned about the true scale of
the amphitheater: it had been about three times larger and higher than
the one we could see, with three tiers of arches and columns behind the
stage, numerous statues, and a seating capacity of twenty thousand.
   Life imagined in Roman times became more vividly alive when we
walked through a North African district, home to Moroccans, Algerians,
and Tunisians, and on visiting the arena. The two rows of columns, one
standing on top of the other and surrounding an arena where gladiators
once fought or chased wild beasts, make the building look even more
imposing than the amphitheater, although simpler in style and easier to
absorb. It is on a hill and dominates the view. By the fourth century CE,
Arles was second only to Rome in size, having taken over from Trier as
the administrative center of Gaul. The arena and the amphitheater together convey the atmosphere of an important center, which is borne
out by artifacts in the antiquities museum (the "blue museum"), from
jewelry to sarcophagi. The hull of a boat that had been found and raised

from the Rhône nearby occupies the center of the museum. Busts of emperors reminded us that Hadrian was a brute, in contrast to the more thoughtful Tiberius.

The reason we were in Arles was to give the tenth annual Heinz Hafner Lecture at the Tour du Valat research station in the Camargue. Luc Hoffmann had established the station some fifty years earlier for the conservation of wildlife and the way of human life of the Camargue. A morning trip to the Alpilles mountain range provided an experience of walking among aromatic Mediterranean shrubs in open, white limestone country, in superbly clear light, and watching a pair of Bonelli's Eagles flying lazily above us across the wide space between two small hilltops. We strolled down a country lane and looked at black bulls on the other side of a ditch and a fence of barbed wire. They are raised for fighting, locally and in Spain. I joked with Rosemary when they ran away that she had scared them (the truth is the opposite). The day after we returned to the United States, we learned from the BBC News that somewhere not far away from that spot, an angry bull had left the herd, broken through the fence, and gored a German tourist to death. No joke indeed.

That evening in the Camargue was special; I should say unique. Together with about a dozen others, we were guests of Luc Hoffmann in his daughter's beautiful, modern, high-quality restaurant, out in the country at Le Sambuc. Luc Hoffmann was there, remarkably strong at ninety, despite needing the support of two walking sticks, so was his large and lively son André, head of the Hoffmann Foundation, and an even livelier administrative assistant, Louise. The really special guests were members of the Arles Van Gogh foundation. I had the good fortune to sit next to Theo Van Gogh's great-grandson and across the table from his sister and had a most enjoyable evening talking, inevitably, about Vincent Van Gogh, his life, and his paintings. Both Mr. Van Gogh and his art expert companion Mr. Van Heughten told me that Vincent's best painting was one of those known as *Sous-bois* (Undergrowth), so I made a note to see it. I don't know whether they thought, as I do, that Vincent chose to paint tangled roots because they mirrored his mental state at that time. That conversation set Rosemary and me up for following in the artist's footsteps the next day, to various places he visited and painted, to see what he saw and chose to represent: the yellow café at the Place du Forum, the stars at night from near the Trinquetaille bridge, the entrance to the Arles public Gardens, the Alyscamps (Roman necropolis), and the hospital where he stayed in 1889, preserved in its original state. Only recently has the location of the particular *Sous-bois* painting been identified.

We returned to the region, specifically to Montpellier, in early 2016, invited by Ana Rivero, who had spent a postdoctoral year in Princeton in our department. The weather was far from Mediterranean, and so was the diet. Montpellier has a well preserved medieval center on a hill, including France's oldest medical school and adjoining botanical garden, both originating in the sixteenth century. We spent three days talking to students and faculty, and one evening with the ornithologist Jean-Louis Martin, who I had met at a course at Erken just outside Uppsala in 1979 (chapter 12). His admirable wife, Silvie, gained her PhD in social economics at the venerable age of sixty-five. At the end of the week, we sang for our supper—that is, we gave the Louis Thaler Lecture to a room full to overflowing.

The evening after the lecture, we drove for two hours for a weekend at the Ferme d'Ajas near Compagnac. The "we" in this case was Ana, Sylvain, and their two children, and another family, Karen McComb, Thierry Boulinier, and their two children. The farm accommodation was a converted old *manoir*, with very large rooms and thick walls. The weather improved, and the weekend could not have been more enjoyable. On each day we went to bed late and slept late. On the Saturday we drove to the magnificent Gorges du Tarn, a valley deeply incised in dolomitic limestone, then over the high ground, the *causse*, to visit a cavern called the Aven Armand. To enter we took an underground cable car four hundred feet down into the ground. Rosemary alone knew what to expect because she had seen a postcard illustrating it. The view of the cave was unbelievably breathtaking, otherwordly. I was speechless, and maybe breathless, incredulous. It was enormous and full of majestic stalagmites. To give an idea of scale, Notre-Dame cathedral could fit inside! The highest stalagmite stood a staggering hundred feet tall. The columns of calcium appear to have been decorated naturally at intervals with the same flared leaf design as the capitals of Greek Corinthian columns. Others are perfectly smooth all the way up. Yet others close to the walls have the appearance of coppiced trees or of the gills of mushrooms. They are floodlit, and the effect is enchanting, a fairyland. Stalactites are utterly insignificant in comparison with the stalagmites. A zigzag path leads to the bottom of the cavern, where the temperature is constant at 54°F. We retraced our steps to the top, pausing to look up at the small hole in the earth above us that Louis Armand had first entered by rope.

The next day we returned to the general vicinity, to a wide valley with exposed rock faces capped by imposing blocks of limestone, and for the very different purpose of looking at vultures. Plenty of them, Griffons, were floating in the sky. With Olivier Duriez as our guide, we walked through pines and oaks, much like the habitat below Jeizinen in the Rhône valley. Orange-tip Butterflies signaled springtime. Eventually we reached a clearing and lookout from where we could see the vulture chicks at nests on ledges below overhangs across the valley. There are 540 Griffon Vulture pairs in France, thanks to protection and food provisioning. At another lookout we watched more than one hundred vultures fly in from a blue sky after Olivier exposed a goat carcass at a feeding site. They squabbled as they wrestled the carcass for the innards with their vacuum-cleaner-like rubbery necks, until the first of five monstrous European Black Vultures glided down, stretched its wings wide, and muscled its way in, with the gait, swagger and authority of John Wayne. We were lucky to see a single Egyptian Vulture, which feeds on small prey, but missed a rare fourth known to the area, the Bearded Vulture.

On the last day of May 2011, we flew to Europe for a combination of lectures (Porto, Seville, Möggingen, and Munich), teaching (Guarda), and a hiking holiday (Jeizinen). In Porto we gave a lecture at the Serralves Foundation. The country was €78 billion in debt, and a socialist party was voted out of power on our second day, but although the mood among scientists was understandably gloomy, our three hosts—Paulo Alexandrino, Paulo Alves, and Nuno Ferrand—showed us great hospitality. Paulo Alves took us to see a megalithic site on the banks of the Douro River. Viewed from the cliffs above, under black skies with menacing thunder in the background, the scene evoked a hostile and threatening environment for our ancestors here more than twenty thousand years ago. I wish we had been able to take a closer look at the encampment in order to visualize the ancestral life better.

In Granada we climbed through the Arab quarter, Albaicín, for a marvelous view in beautiful light of the Alhambra across a small valley. In Arabic, *Alhambra* means "the red house." We learned the significance of the name on a different occasion when staying at the Carmen de la Victoria, a hotel owned by the University of Granada. On the evening of our jet-lagged arrival, we had been dazzled by a magnificent view of the Alhambra from the terrace as the sun went down and the structure turned a golden red.

A chance encounter with members of a math and physics conference gave us the unique opportunity to visit the Alhambra with them at ten o'clock one evening. It was a precious experience to be there in darkness, under a half moon, with a handful of lucky tourists, and quietness. At one point, all we could hear was the trickling of water into the pool of one of the courtyards and nothing else. Rosemary and I strolled in silent awe from one room, with intricately carved wooden-paneled walls, through sculpted arches to another, imaginations unrestrained by noise and senses heightened by nighttime perfumes of plants. With our children we once had a similar privileged experience at Machu Picchu in Peru, when we were allowed to walk all over the site at sunrise without any other people. Nothing could be better for transporting the mind to a culture remote in time, with undisturbed freedom to imagine another life, another world, in precisely those surroundings.

From Granada we traveled by train to Seville, where the next three days unfolded according to a well-exercised routine of combining work with pleasure. The pleasure was combining exploration of the cathedral, the Alcázar from the Moorish days, and the Archivo General de Indias with a trip by car into the country with our host, Jordi Bascompte, and fellow ecologist Miguel Fortuna. Taking advantage of its proximity, we visited Córdoba to see La Mezquita, the Moorish mosque in the old town. It lived up to expectations gleaned from pictures and descriptions of row upon row of thick pillars and bicolored, rounded Romanesque arches. Built in the eighth century, it was taken over by the Catholic church in the thirteenth century, which imposed its domination without risking outright rebellion by building a cathedral tower right in the center and decorating the side chapels, resulting in one building for two religions. A few decades later, an enlightened Catholic king, Alfonso X of Castile, ruled with a scholarly interest in Arab, Jewish, and Catholic cultures. Then the light went out: the long period known as *La Convivencia* (coexistence) ended in 1492, with the capture of Muslim Granada by the powerful Catholics from Castile and Aragon and expulsion or forced conversion of Muslims and Jews.

These travels were briefly halted by hip replacements. My left hip had been replaced with a titanium one by Dr. David Mattingly in the New England Baptist Hospital in 2004 (chapter 17), and the right one was due for the same. The earlier operation and my recovery in the following four days had gone extremely well, and the nursing care and physiotherapy in the hospital could not have been better. Well, there was one exception. One night an intern changed the bandage over the incision. "This will hurt only momentarily," he said, as he ripped it off in one stroke. Ten minutes later I had to call the nurse because the pain was so intense. She removed his bandage. "Oh dear," she said. "That's the problem. All the skin has gone!" In our polite conversation beforehand, the intern had told me he had been an undergraduate at Princeton, and I decided I must have taught him in Introductory Biology. Several of us had often joked about the ultimate pre-med student's revenge on the professor for not giving him an A grade in class, and here I was experiencing something like it! As for the new hip, recovering 90 percent of full mobility at home was relatively easy, and I graduated early from crutches to a walking stick, but getting the rest was a long, drawn-out slog (or drag).

Undeterred, I returned for a replacement of the right hip ten years later. It was done in an hour and was apparently straightforward, according to Dr. Mattingly, who told me that the right hip was an exact mirror image of the left one; they are not always so. I had oxygen in the nostrils all day and felt beautifully rested, happy, a few degrees short of delirium.

My only moderately sensible thought that day was the question, What will peoples' moods be like in a future with rising atmospheric CO_2 and lowered oxygen? At about two-hour intervals I had to repeat my name and date of birth when prompted by a nurse or intern: Peter R. Grant, 10/26/36. When I awoke the next morning, I found they had formed the basis of a poem about my incarceration in this hospital (below). Back in Princeton I transferred to physiotherapy at the Neurac Institute, which specializes in exercising patients partly suspended in midair in slings, so that the individual leg muscles can be exercised while free from supporting the body, a system called the Redcord method. Big Nick Passe was my genial therapist. In the following year Rosemary and I traveled by train to Boston for my checkup. Dr. Mattingly took about three minutes to tell me I had done very well. I told him *he* had done very well. Years later I am still hoping that is true for the long-term.

Hip Replacement

It looks like a fortress on top of the hill,
has an air of impregnable strength.
And once you are in, you are in, **you are in**;
and only they say when you get out.

Everything is done with courtesy and grace,
and any thought of running away,
is out of the question once the plastic ID
is handcuffed to your wrist to stay.

"Name; date of birth?" you are asked at the desk.
You will hear it twenty more times.
So please don't forget that month comes before day:
ten; twenty-six; thirty-six.

Undressed and wired up, you are wished all the best
by your wife as they wheel you away.
You've not gone very far, when strange to relate,
you disappear into thin air!

Equally strange you come back from elsewhere,
but now you are in a small cell,
curtained off by plastic from the world outside.
Have I lost my sense of smell?

Your brain has shut down, or taken a break,
and floats somewhere up there in space.
Yet before they give you the pills you must say:
ten; twenty-six; thirty-six.

Did she say I will be given some broth,
and try to stand up just once?
And there is the surgeon who says all was fine.
Any questions? I mumble my thanks.

I would have slept throughout the night
if my neighbor had not loved TV.
And the next night too if he had not been sick
at eleven; at one; and at three.

So, I went into training to get out of that place
at constant 70 degrees.
To do so I had to pass rigorous tests
from staff who were called PTs.

A reasonable guess is that PT might stand
for some form of Physical Torture.
But Physical Therapy was what they gave,
and I passed their tests in good measure.

To do so meant I had to walk every day,
with a crutch on my left and my right.
While stealing a glance at those others in cells,
at the risk of a frightening sight.

After three days they did indeed let me out,
propped up on my crutchiform sticks.
But only after I had recited once more:
ten; twenty-six; thirty-six.

I haven't escaped, I'm just on parole,
reporting to another therapist.
Big Nick puts me in slings, and suspended in air,
I'm forced to reveal my weakness.

The punishment I get is a daily stretch
of my quads, my glutes, and QL.
And slowly it seems increasingly less
like a prescription they sent from hell.

My hangman who pulls on the red-corded rope,
mixes politeness with an occasional frown.
But, he insists, when you are up, you are up;
and only I say when you come down.

The company Neurac is Norwegian for Nerve Wrack,
a nerve wreck is perhaps their goal.
The Spanish Inquisition is what's on my mind,
the rack where body leaves soul.

OK, I am joking, it's doing me good,
and my muscles are gaining in strength.
Soon I'll be free to wander the world over
with hamstrings at their proper length.

Or so I thought when I left friends at Neurac
like a bird taking off in flight.
But with exercise ordered at least twice a day,
normal life is not yet in sight.

23

Hobbling in Hawaii
and Becoming an American

ON APRIL 2, 2014, our book on the forty-year study of finches was published, to no fanfare; that came later: "Its value lies not only in the scientific conclusions drawn from the data, but also in the insights that this
book gives regarding the value of long-term field studies" (Webster
2015, 2247). (For a poetic version, see the Appendix.) In June we began
the large, long, and enjoyably creative task of making seven short films of the
Daphne research. Several helping hands in the New Media Center lab on
campus guided us novices. Lisa Jackson enthused as she showed us how
to adjust sound and make media titles and credits and gave us this pearl
of wisdom: "A project is never completed; it's only abandoned!" In the
previous year we were filmed with Sean Carroll as part of his educational
effort to explain and illustrate evolution for the Howard Hughes Medical
Institute. Filming took place in a lab in Guyot Hall for eleven hours in one
day. It would have taken 111 hours if I had not said "enough" to the give-
an-inch-take-a-mile producer. The product was very good, much better
than we feared, and professionally produced but still left plenty of scope
and an abundance of motivation to produce our own personal account of
research on Daphne. After more than a year we "abandoned" our project,
signed contracts, and delivered videos to book publisher Pearson Education and to Facts on File of the Films Media Group to use on YouTube.

Pearson gave us an inside view of its world by inviting us to the Biology Leadership Community conference for high school and college

teachers it sponsors every year. Neil Campbell had started this venture sixteen years beforehand, and it has continued every year since, in part as homage to Neil, who died after a couple of years at fifty-eight years of age, the same tragically early age as my father. The meeting was in Pearson's building outside Austin, Texas. The teachers made short presentations or discussed topics in workshops. We were very impressed by these people who were dedicating their lives primarily to teaching and only secondarily, if at all, to research. For me, the high point was a conversation with Jamie Jensen from Brigham Young University, who described an exercise for teaching the essence of making evolutionary inferences about the past that involved placing a complex Lego structure in a bag and having students feel the hidden object and then make a copy or model of it with individual Lego pieces. The structure was too complex to be successfully copied. Students could return, re-feel, and revise the copy. They got the powerful message of uncertainty about the past and the necessity of using partial information to estimate it.

In the evening after the conference, we watched a half million Mexican Free-tailed Bats streaming out of their roost beneath a bridge over the Colorado River. The stream was continuous but never congested, as if someone were controlling the flow. Like a flock of thousands of starlings or queleas, or smoke in the wind, a flock of bats coiled upward through the air and away to the Gulf Coast for a night of insect feeding. We returned the next night for an encore and decided there were twice as many because the exodus took twice as long.

The first cicada (*Tibicen lyricen*) sang on July 3, and a few days later we were off to Hawaii to attend an island-biology conference in Oahu. It was a pleasant gathering of about four hundred people, small enough for the atmosphere to be very friendly, and a welcome opportunity to see old friends and evolutionary biologists like Francis Howarth, Sheila Conant, and Bob Ricklefs, and to get to know a few others. We were bused each day from the New Otani hotel at the end of Waikiki Beach to the university. Common Mynas fed among picnickers on the lawn, and Fairy Terns flew like wraiths among the coconut palms and casuarinas. The hotel is distinguished by a couple of old, spring-watered hibiscus trees at the edge of the beach where Robert Louis Stevenson once sojourned and wrote.

Maui, after the conference, was a new island for us. On day one we drove a rented car up to the highlands, to Haleakala. Then, with light packs and, for me, a Leki walking stick, we climbed down into the crater in bright sunlight and were rewarded with views of both spectacular moonscape scenery and the endemic silverswords in flower. Silverswords are named for their leaves. They are the Darwin's finches of the plant world, having diversified into many species, and like them they occasionally hybridize. At a split-rock end point, a young man told us: "You should really go down to the bottom of the crater because the silverswords are much better there." So, throwing caution to the wind, we did, and his advice was fully justified by the silverswords in various stages of expansion to full flowering, mostly scattered across the granular, cindery soil, some lining either side of the path, like an avenue or guard of honor. The flowers hung down, waiting to be pollinated by small bees and flies. We paid for the privilege of seeing them with the rigors of climbing back to the top, from 1,650 feet to 2,300 feet elevation. First the mist that spread downward in wisps turned to light rain and then to a downpour. As the slope became steeper, the lengths of our climb between successive rests became shorter. Limb and Leki supported me well, however.

We had asked Hanna Mounce, who was studying the famous Hawaiian honeycreepers, if we could visit Waikamoi nature reserve on the wet side of the island, and she very kindly agreed to take us there. I had paid for this in advance by giving her some colored leg bands for the Maui Parrotbills (or *kiwikiu*) she was studying. Walking in the rain forest was a great experience. In mixed disturbed forest we had a wonderful view of the creeper known as the Maui Alauahio, which looked like a North American Blue-winged Warbler. Then, in the undisturbed forest, we saw more of the creepers, one *apapane*, one *iiwi*, and, new to us, two or three Crested Honeycreepers (*akohekohe*). The closest we came to a parrotbill was following a call down the trail. Apparently they are often detected by the crunching sound they make while searching twigs and small branches for cryptic insects. They break a twig by separating their mandibles horizontally (laterally), then applying pressure vertically. We did not hear the telltale sound; at least I didn't. The parrotbill is an interesting ecological counterpart to the Woodpecker Finch in Galápagos, which has solved the problem of reaching insect larvae deeply hidden in woody tissues differently, by using a twig as a tool to extract them. Throughout the walk, our heads were in misty rain, and sometimes heavier rain, so we were fortunate when the Cresteds with their striking crests came into view for a few seconds before disappearing.

The experience was a potent demonstration of the extreme difficulty of working with these birds. We learned that the forest was split into two by a deep gulley, and that approximately four hundred parrotbills lived in one part and one hundred in the part we were in. They differ genetically in the two parts, and the Crested populations differ even more. Once connected at lower elevations, the two parts of each population can no longer make contact because avian malaria has set a lower limit to their distribution at forty-five hundred feet, which is above the point where the gulley ends and the forest segments are united. The honeycreepers are a sad example of a once spectacular adaptive radiation, three times more extensive than the Darwin's finch radiation, older, and much more colorful, but unlike the intact Darwin's finches, decimated by habitat destruction and introduced disease.

One evening we were at Nicola's house in Corvallis discussing an impending visit to John Bonner in his ashram. "What is an ashram?" Anjali asked. "A place where people retire," I replied. Nicola added a joke: "We are going to put grandmummy and granddaddy in the ashram." Anjali: "Why, because they don't know how to retire?"

Well, actually, yes.

We spent the fall semester of 2010 at Yale University in New Haven on visiting professorships. Just as going to UBC a couple of years before was a return to our graduate days and environment, so was this a return to the postdoctoral phase of our life, to the city where Nicola was born. This time we stayed in 325 Audubon Court, conveniently placed a couple of blocks down Whitney Avenue and Church Street from the

Peabody Museum and the Environmental Sciences building. Our previous address, 558 Whitney Avenue, looked exactly the same as in 1964, whereas the food shop we frequented on Orange Street had disappeared, and even its location was difficult to identify. We justified our existence by teaching a graduate-level course. Teaching may not be quite the right word for eating pizza and discussing a wide range of aspects of speciation with one to two dozen graduates and postdocs for one hour a week in Osborn Hall, room 400.

After an initial settling-in and adjustment period, we became thoroughly in tune with our life at Yale. One large factor was the music. Thanks to generous alumni benefaction, much of the excellent music abundantly on offer is free. We attended chamber and full orchestra concerts, performance on baroque instruments, a baroque opera (*Scipione Affricano*) sung by undergraduates, and, best of all, full-screen, high-definition, live broadcasts of opera from the Metropolitan Opera House in New York on Saturday afternoons. *Boris Godunov* followed *Das Rheingold*; then came *Don Pasquale* and finally *Don Carlos*. The singing was magnificent, and the transmission was almost always impeccable. Both the acting and the staging have improved enormously since we first went to operas in the 1950s. In the intermission, we could walk home and use our own bathroom if the concert hall's one was too crowded and not miss the next act.

I have vivid memories of my trip to the Everglades in 1959, and some of these came back on my next visit, fifty-seven years later, when Rosemary and I were invited by Al Uy to be visiting professors at the University of Miami for two weeks. Our interests were similar, as Al has studied the evolution of bird populations on islands, including the Galápagos. After a week of numerous discussions and interactions with graduate students and a few faculty members, we were taken to the Everglades by Bill Searcy and a dozen students. Despite light rain, the Everglades experience was highly stimulating because the environment is unique. While our eyes were directed mainly at the abundant waterbirds, someone pointed out an alligator, 99 percent submerged, identical to an image in my memory, and then suddenly my memory received a jolt as I noticed we were walking on the Anhinga Trail, just as I had in 1959. The hammocks of the Everglades, little islands of vegetation surrounded by marsh and water, are especially interesting ecologically. Their value lies not only in their composition but in their provision of food and safety for migrating birds. One, the Mahogany Hammock, is a mini subtropical forest, complete with a red-barked *Bursera* tree, a link with both the Tres Marías islands and the Galápagos. The substrate is only a few inches above the level of the rest of the marsh, which in turn is only three to four feet above mean sea level. In a world of rising sea level these precious habitats are precarious.

For many years, seminar invitations came to both of us, and each of us contributed half (and still do). Increasingly, Rosemary received an invitation alone, partly the result of her growing reputation as a public speaker and partly owing to a shortage of senior women. A special event was the Lady Margaret Lecture, given to an audience of about one hundred at Christ's College, Cambridge, sponsored in part by the Galápagos Conservation Trust. Lady Margaret Beaufort was a remarkable woman. She founded the Tudor dynasty. At the age of thirteen she gave birth to the future Henry VII, father of the infamous Henry VIII, and with that her reproductive life came to an end. (Many would say that was more than enough.) She founded Christ's College (motto: *Souvent me souviens*; "I often remember," or better, "Think of me often"), Saint John's College (same motto) and a girls' school. We stayed in the Master's Lodge in Lady Margaret's room, following in her footsteps, from more than six hundred years ago, up a stone staircase. A small recess behind the room allowed her (unseen) to observe the dining students below. Comfortably modern, the room nevertheless had an ancient, chain-operated water closet above the loo, just as I remember the Department of Zoology did when I was a student. I expect it froze in olden times or else was kept active with heat from a fire. The master, Frank Kelly, and his wife, Jackie, were very warm, welcoming, and unfussy hosts.

The occasion was made especially enjoyable by the opportunity to chat with numerous Cambridge friends over the buffet dinner and wine afterward. Roger Perry, past director of the Charles Darwin Research

Station, had traveled from Suffolk, and among nuggets of information about the Galápagos, he recalled Karl Angermeyer, one of the early settlers on the islands, telling him that the Wittmers, other early settlers, must have dispatched and buried two unwelcome visitors to their island paradise of Floreana, an outrageous self-styled baroness and her male companion, named Phillipson. This happened close to one hundred years after Darwin's visit to the same island. I remember Karl hinting at the same to us. The story was told entertainingly in *The Galapagos Affair*, as a mystery without a solution, by John Treherne (1983).

Rosemary manages to move some young women almost to tears when she gives a lecture on Darwin's finches. She did so at the International Society for Behavioral Ecology Congress on the attractive Exeter University campus in 2016. Her lecture received prolonged applause and a partial standing ovation. I could not hear talks very well, apart from the plenaries, so I decided to spend afternoons in Exeter in and around the splendid twelfth- to fourteenth-century cathedral. That and the signs of Roman occupation made the walks unexpectedly interesting. We stayed at the Royal Clarence Hotel, which I mention because it is reputedly the oldest hotel qua hotel (not an inn) in Britain. It was built in 1759 on the edge of the green in front of the cathedral. Our bedroom was constructed around an enormous bath that was presumably too heavy and too large to move. Sadly, the hotel was destroyed by a fire in the following October. I have no doubt the bath survived.

A lecture in Helsinki is a third example of Rosemary going solo. The Finnish Society of Science and Letters in collaboration with the younger Finnish Academy of Science and Letters hosted an afternoon symposium in honor of the ecologist Ilkka Hanski, who had died of cancer in the previous year. It was 2017, the one hundredth anniversary of the independence of Finland, and the symposium was the first of four to mark the historic occasion. Rosemary, Jared Diamond, and Dieter Ebert were the only speakers. We knew Ilkka and admired his work on islands and his ability to communicate it (Hanski 2016), so Rosemary put a lot of effort into a well-crafted eulogy and description of Ilkka's research, interwoven with our own research because they were so similar.

There followed a seminar visit to Jyväskyla, Finland, thanks to an invitation from Johanna (Jonna) Mappes, a biology professor who specializes in behavioral ecology. The next day was devoted to the odd combination of bird-watching and looking at six-thousand-year-old rock paintings—the fruits of our joint labors. With Jonna and about a half dozen wizard birders, we found a Ural Owl and then a Great Gray Owl in separate locations in thick forest, both birds motionless, like cardboard cutouts against a dark background. Leaving the forest, we visited a mire, a wet sedge habitat atop several yards of peat sparsely populated with short pine trees. A Whinchat was notable in being the only bird anyone saw. Returning to the dark coniferous forest in order to visit the rock paintings at Saraakallio, we walked for a third of a mile, then climbed down a steep cliff to the shore of a lake, so that we could then look up at figures painted in iron-ore red on pale vertical rock surfaces well above us. Some were difficult to identify; others were clear: men in a boat, moose and deer. They were quite unlike the spiritual figures we have seen elsewhere (chapter 24), representational rather than mystical, unless I am mistaken. Presumably they were painted when the water level was much higher or, perhaps more likely, when the ice was at that level.

We had been meaning to apply for US citizenship for many years, but it was not urgent, and we continued to postpone it until a discussion about inheritance tax tipped us into action. Here we had the luck of help from Dan Berger, an immigration lawyer, and husband of an erstwhile Princeton colleague and friend, Laura Katz. On October 15, 2014, we drove to Mount Laurel, New Jersey, for an interview with the US Citizenship and Immigration Services.

There had been a long and uncertain lead-up to this interview, with Rosemary's appointment first being scheduled for August, then for September, and then mine for October 15 at 8:30 a.m., and finally Rosemary's added at 9:00 a.m. In view of all this, it should have been no surprise that Rosemary was called in for an interview first. Everything went smoothly until I was told that my permanent file had not been sent from the archives in California, and, unlike Rosemary, I could not take the oath of allegiance that afternoon. I would have to wait a couple of weeks or more. Rosemary decided to postpone hers to coincide with mine. Back home, and while we were rearranging our plans after lunch, a phone call from Rosemary's interviewer told us that my application had been approved, and we could take the oath tomorrow. Surprise, surprise! So, we hurriedly made an appointment by phone to submit applications for US passports to the Philadelphia passport office at 10:00 a.m. on October 17.

Back we went to Mount Laurel the following afternoon, all dressed up for the ceremony at three o'clock. Before going into the building, I managed to leave all my documents in the Starbucks' loo. As I left, a desperate man shot in and locked the door behind him. Rosemary distanced herself and read every label on the bags of tea and coffee on the shelves while we waited. Eventually the man emerged, after five exceptionally long minutes, and apologized; I reclaimed my folder, and thereafter we were back on track.

Thirty new Americans from sixteen countries took the oath. India was the majority country, England a minority. I sat next to a large Jamaican woman dressed in bright orange while her two- or three-year-old hyperactive son ran up and down and occasionally screamed. For her part, she was really into the ceremony, very excited whenever the words "liberty" or "freedom" were mentioned, saying, "Yeah!" The director in charge of the ceremony performed his role very well, setting just the right tone of seriousness and enjoyment. I think he must have exercised some high-level power to override the rule that approval for citizenship can be given only when the permanent file is present. We chose the time and the US president well: Barack Obama gave a heart-warming speech on video. The ceremony lasted for exactly one hour.

We slept deeply that night and were up early the next morning to take the train to Philadelphia. Once again, everything went smoothly with our applications for passports, until right at the end when we were told they could not be issued that day because we were not traveling (to Canada) for another week or so. "Please come back at ten a.m. on Tuesday," four days hence. When the time came, I erred by writing "brown" in the box for hair color. When I discovered it, I thought it was so funny I told Rosemary: big mistake. "You did what?" she almost screamed, laser-like eyes boring two holes through my cranium. "You are an idiot. You haven't had a brown hair in your head for twenty years!" I could visualize the wheels whizzing round in her worried brain: *Lying and failure to report correct information on a government document is subject to a large fine and long jail sentence for US citizens, and immediate deportation for non-citizens.* I crossed out "brown" and put "white," although I wanted to put "gray" to soften the transition a little. In reviewing our applications, the woman behind the grille at the counter batted not an eyelid as she read: height 5'11", hair ~~brown~~ white, eyes blue, sex male. Perhaps she thought, *The country needs more white-haired old men, and anyway he is going to leave the country after a week.* A return trip to Philadelphia five days later to obtain our passports ended the business.

The issue of citizenship came to the fore in a different way in December. Robbert Dijkgraaf, director of Princeton's Institute for Advanced Study, and his wife, Pia, hosted two fellows and officers from the Royal Society and about fifteen Princeton fellows of the society at his house on Olden Lane for dinner. The visitors from the United Kingdom were Tony Cheetham, in charge of the Royal Society's US relations, and John Pethica, the society's physical sciences secretary. The most interesting part of what they had to say was a potential move to allow British people who take out US citizenship to be eligible for election to the Royal Society. I made a pitch in support of the move to Tony after the speeches. Like me, he felt strongly about it, having taken out US citizenship after becoming a fellow and before returning to the United Kingdom. Most of us British-born at the dinner were dual citizens, and all for the same fiscal reason.

On December 16, we had the most contentious faculty meeting since the unruly days before the EEB was formed as a separate department. I left feeling that I had overstepped the emeritus professor mark by speaking too much and vowed not to speak at the next meeting. We came home to find an e-mail message from the Arnside Cemetery informing us that a space in the Matchett family grave was ready for both Rosemary and me.

Did someone in official channels find out about that? A few months later, we returned home from a seminar visit to discover a letter from the US Social Security Administration addressed to the family of Peter R. Grant, which stated that according to their records, Peter R. Grant was deceased and requested information on recipients of benefits. Like the rumor of Mark Twain's death, the report was greatly exaggerated and indeed highly suspicious of attempted malfeasance. We had to get help from our excellent lawyer, Fritz Cammerzell, with a letter to Social Security, complete with a photograph of me holding the day's *New York Times*. We never heard whether I had been officially undeceased or ceased to be deceased; however, Social Security income continued, and I continue to pay tax on it.

24

Guests at the Top End

ONCE I HAD the experience of visiting a region where not a single species of bird was familiar to me. This was New Guinea and Australia in 1974 (chapter 9). Ecological roles were performed by species that looked similar to familiar ones but were in fact quite different in evolutionary origin. Contrariwise, some species seemed to have stepped outside the familiar and entered new ecological niches, like the kingfishers, strictly riverine in Europe and North America but there with relatives that are forest-dwelling insectivores or crow-like predators of lizards and snakes (e.g., kookaburras). And then there are those with no counterparts in the Northern Hemisphere, like lyrebirds, whipbirds, and logrunners. Collectively, the fits and the misfits, with the naturalist's prior experience, stimulate him or her to wonder about the roles of time, ecological opportunity, history, and geography in forging the characteristics of a community of birds (Price 2008) or any other group of organisms (Darwin 1859). They stimulate you to think differently.

I returned to Australia in 2016, with Rosemary, and this time it was her turn to witness entirely new ecosystems. We had the good fortune and privilege of being invited to Australia to spend the first three weeks of June as the 2016 Charles Darwin Scholars at Charles Darwin University in Darwin, Northern Territory, at the "Top End" of Australia, latitude 12° south. We stayed on campus in an apartment in International House that had everything we needed. On the day after we arrived, the weather tested us when the maximum temperature (97°F) broke the June record for Darwin. Sleeping was difficult at first, with the overnight minimum of 81°F outdoors and higher in our apartment. After a week or so, the temperatures fell comfortably to 90°/72°F, but it was dry, the sun was very strong, and I have never drunk so much water while at the same time remaining almost permanently on the edge of dehydration.

Bird-watching began early, while we were having breakfast on the balcony or on walks to the beach about a half mile away through a small patch of mixed mangrove woodland. As viewed from the balcony, our breakfast birds were four Bush Stone-curlews, utterly motionless, like sculptures, two mound-building and scratching megapodes, ibises, Masked Lapwings (who we dubbed the "parking lot gang"), honeyeaters, Magpie-larks, and Double-barred Finches. Rainbow Bee-eaters were the most colorful. Black Kites were everywhere, the only species among 130-plus we saw in Australia that we had seen in the Northern Hemisphere, if the recent splitting of the osprey into two species is justified. Each evening, for about twenty minutes, flying foxes (fruit bats) streamed out of their roost in the mangroves to various feeding locations. By seven o'clock the exodus was all over. I counted 352 one night.

Our duties were light: to give a public lecture, called an oration (Fig. 24.1), and a departmental lecture at a slightly higher level; to engage in the making of videos on evolution for a massive open online course (MOOC) by responding to questions from biologist Keith Christian; and to meet many people and discuss research with them. We were interviewed for radio at the Australian Broadcasting Corporation building in downtown Darwin; by a graduate on campus for an article in the student magazine *Flycatcher*; and by a group of Rotarians over dinner.

The university chancellor, Simon Maddocks, had invited us. He spent his early life growing up in Papua New Guinea and had had a distinguished career as an animal and agricultural scientist before taking up his position—an admirable round peg in a round hole, as my father would say. Maryanne McKaige, the Charles Darwin Scholar program coordinator, was equally impressive in her role and never failed to ensure that our program ran smoothly. Stephen Garnett and Keith Christian acted as naturalist advisers. Keith took us one early morning along a boardwalk through a rich mangrove forest a couple of miles from campus and gave us one of our ornithological highlights by using playback

FIGURE 24.1. Meeting Charles Darwin at Darwin University, Northern Territory, Australia, June 2016. (Photo Simon Maddocks.)

from an app to attract two Rainbow Pittas, built like European Robins and colored like birds-of-paradise—exotic!

A great benefit of the program was the availability of a Hyundai sports-utility vehicle on weekends. For the first, we joined a group of naturalists in Litchfield National Park and camped in a borrowed tent. Unfortunately, a planned mammal-trapping program had to be canceled because too many fires were being lit in the national park's controlled-burning program. Instead, we watched as awful Cane Toads came out of the bushes to visit us at the campsite. They were introduced to Australia to control beetle pests of sugarcane and became pests themselves by being toxically protected from would-be predators, eating a great array of insects, frogs, and snakes, and reproducing prolifically. The situation is interestingly dynamic as the first signs of evolution have been detected, both in the toads (faster dispersal) and predatory snakes (greater tolerance of the toxin). More interesting to us as observers was the remarkable rocket frog, as tiny as one phalange of my thumb yet capable of jumping six feet.

We became fascinated by the landscape of termite mounds made by several species. The tallest, called cathedral mounds, reach nearly twenty feet. The most biologically spectacular are the magnetic termite mounds, which are oriented north–south. Flat-sided, like gravestones, they stand out in the open on seasonally flooded fields of grass growing from a viscous black mud, rising up to six feet or more above the ground. To withstand the seasonal flooding, the termites store their food in chambers above the high-water mark. The mystery of the mounds is how a new one becomes established and grows fast enough to escape total immersion. Stephen Garnett has studied them and told us there is a graded age from bottom to top as expected, and that he has evidence the grains at the bottom were exposed to light two thousand years ago. Perhaps the mounds were initially constructed in a past period of dryness. Sadly, the mounds are being slowly eroded by feral cattle and pigs.

Kakadu National Park has a different attraction, in the form of rock art, an abundance of it under various overhangs of rocky sandstone strata. We visited four major sites and were impressed by how much can be seen, how well preserved it is, and also by its age, twenty thousand years in some cases and at one of the sites possibly much closer to the arrival of the Aboriginal people some fifty thousand years ago. The spindly human figures contrast with the larger, very full-bodied depictions of fish, principally the Barramundi. Some of the illustrations are spirits in human form; some reflect human activities; others tell a story with a moral concerning "what thou shalt not do otherwise thou wilt suffer."

In one story, two sisters turned into crocodiles in order to eat humans. These rock paintings of strange figures connect the physical and spiritual worlds in many ways (Clottes 2016), like the diver into the underworld at Paestum (chapter 22) but more fearfully.

The paintings were the focus of our walking activity, and we could not have had too much. Being completely alone in the bush was inspiring, although potentially dangerous, as we discovered how easy it is to get lost when straying a short distance from the trail. A Blue-winged Kookaburra and a Great Bowerbird kept us company; we had missed seeing the owner of a bower on campus because it was away on vacation. The best bird-watching was at the euphonious Anbangbang Billabong, a picturesque shallow lagoon festooned with blue water lilies. Crocodiles may have been lurking, but we did not see any, either there or elsewhere. What we did see was a massive boar, so big that we both thought it was a feral buffalo at first sight. It was partly immersed in the mud at the edge of the water, and then suddenly sprang up and frighteningly dashed out and away. On returning to Darwin, we stopped at a small billabong very close to the Arnhem Highway to look for the famously colorful Gouldian Finch, following a tip that some may be there to drink. Did we see one? I don't know. We did see a small, bright green bird at a distance, but I saw no bright colors on the head or chest.

The Litchfield experience led to some bird netting near Rapid Creek, close to the campus. John Rawsthorne—the leader of the naturalists' group—invited us to join him on a Thursday afternoon. I was recovering from a brief and mild flu, so a leisurely afternoon netting with frequent sitting in chairs was ideal. Slim pickings, as we caught only eight honey-eaters and flycatchers in two nets. After sunset, nightjars began calling, very loudly, and then suddenly appeared, swooping into our area. Buoyed up, we drove off, only to be confronted by a double-locked gate, and John had to use his crowbar to extricate us. He is a lawyer; that fits somehow. His last gift before driving us home was to show us where a Tawny Frogmouth, a bird that mimics a broken branch, could almost always be found in a clump of casuarinas . . . though not that night. We walked there along the beach the next day, and as we were about to give up and leave, Rosemary spotted one, close to the trunk of a tree, about ten feet up, looking extraordinarily cryptic. An amazing bird with an enormous mouth, it has evolved independently in the same direction as the broken-branch-mimics of Central America, the potoos. I rank this with the Rainbow Pitta as an ornithological highlight.

A year would have been thirteen times better than a month, but regrettably we could not stay that long. On our last day in Darwin, the United Kingdom, more remote than ever on the other side of the globe, voted to leave the European Union; we watched the results coming in on the BBC News, and like everyone else around us, we could not believe what we were seeing. That morning our hosts gave us a warm farewell at a ceremony in the university; then we flew to Alice Springs, where our purpose was to give the oration again at the Charles Darwin University's small campus there. The coolness of the air when we arrived at the airport at latitude 23° south was a mighty tonic.

Abandoning the usual habit of flying everywhere, we took a bus on a six-hour journey through arid country to Uluru (Ayers Rock). The relatively slow journey gave us an appreciation of the vegetation, distance, and remoteness of the country. The road followed the route taken by the Scottish surveyer John McDouall Stuart when he explored the way to the north coast in several treks in the years 1858–61. These must have been amazing accomplishments of navigation, skill, and survival. Rain was falling lightly when we arrived at Uluru, then a little more heavily in the afternoon. Our mood changed when Michael Misso, director of the Uluru-Kata Tjuta National Park, and his wife kindly drove us all the way around the extraordinarily impressive oval-shaped rock that is Uluru. Water was cascading down in waterfalls—white water in the middle of long black algal stains from top to bottom, similar to what we had seen on Daphne (chapter 21), though on a much larger scale. Apparently, such cascades occur only rarely. They helped us to understand the weird erosional features of incised grooves and strings of pockmarks that look as if they had been gouged out of pliant dough by giant fingers.

Australia is famous for its marsupial mammals, which constitute a large-scale adaptive radiation, but apart from wallabies they were not much in evidence. The remnants of the mammalian fauna live on in places like the Alice Springs Desert Park where they can be seen in arenas under reversed day and night light conditions: bilbies, bettongs, malas, and a quoll. All these are endangered in the wild, owing to predation by introduced foxes and cats, and close to extinction. Soon, we were told, several medium-size marsupial species will be alive only in captivity—a depressing thought.

From Alice Springs we flew to Brisbane, via Sydney, rented a car, and drove through some attractive reconstructed English countryside to O'Reilly's Rainforest Retreat, on the Lamington Plateau at latitude 28° south. I had visited the same place in 1974, immediately after the International Ornithological Congress at Canberra. Returning was an ambition fulfilled, this time with Rosemary, and it did not disappoint. The first four days were beautifully cloud-free as we searched for birds and sought an understanding of the forest, mainly on the Border Trail. Looking *for* birds and *at* trees, and with few interactions with people, was a relaxing way of spending a holiday; however, we had to work hard to find and identify birds in the forest. Seeing more than ten species on one day was a rare achievement. We got better at the rapid use of binoculars, the quick aim and focusing that are necessary in order to identify birds in low light while they are momentarily in sight. Their haunting calls always came from somewhere else.

Ornithologically, the most exotic were a couple of female Paradise Riflebirds, several whipbirds, and logrunners. The whip in the whipbird's song is an unusual sound for birds, loud, liquid, and explosive. We finally saw a whipbird close enough to see how it produces the sound by a sudden contraction of the extended neck, so that the head is pulled down into the body and the vocal tract presumably rapidly shortened. These are cryptic birds, an enjoyable challenge to find when they are scratching quietly in the litter beneath dense ferns and other plants. Lyrebirds provide a similar challenge; Rosemary could appreciate their continuously changing vocabulary better than I could, but although we heard them practically every day, we did not see even one. A lyrebird display site near Python Rock may have been the same one I saw forty-two years before. That, the main O'Reilly's building, and the adjacent feeding station for bowerbirds and rosellas were all that I could remember. However, I did remember very well the sharp transition between rain forest and wet sclerophyll (eucalypt) forest, which was so striking in contrast to the usually gradual change between neighboring habitats. It was probably just above Python Rock, where the radiation and wind pattern change rapidly with exposure.

When we were not scanning the forest for birds we stared at, admired, and wondered about the giant trees of the forest, especially brush boxes (*Lophostemon*) and southern beeches (*Nothofagus*), interspersed with trees known collectively as Chilean pines (*Araucaria*). They are all impressive on account of their ages, numbers, and sizes. I read that the roots of one tree had been given an estimated age of five thousand years. They are labeled as remants of the ancient Gondwanaland forests when, more than twenty million years ago, Australia was connected to Antarctica. Walking in this forest is a walk into history even further back in time, for Gondwanaland is where passerine birds originated about fifty million years ago, spreading around the world and becoming sparrows, starlings, crows—and Darwin's finches.

Apparently *Nothofagus* no longer reproduces by seeds because of the cool temperature and lack of light, but they do reproduce clonally, and some of the trees are not really separate individuals so much as clusters of the same one. I kept thinking of a long, drawn-out battle between vines and the trees, and between vines and other vines, an imperceptibly slow hide-and-seek dance between vines as predators and trees as prey—*La Danse Macabre* of the forest. Once a tree is found the battle is clearly won by strangler figs, but how long does it take them to kill their supporting hosts, we wondered, and how long can they persist further on skeletal trees? And why do some trees get smothered, and others escape? Are they chemically protected in some way, or is it a matter of luck? The drama of the giants must unfold over centuries, since some of the *Nothofagus* are estimated to be fifteen hundred years old. As we stared upward in this cathedral of trees, Rosemary lowered her voice in unthinking respect, which was understandable but of no help to me!

At the edge of the forest, misty rain and clouds rise upslope from the valley below. The forest receives up to six and a half feet of rain each year and is probably the most southerly subtropical rain forest. By visiting in the dry and cool season we missed the heavy rain, the birds breeding, and the pittas (which had moved downslope in June), but we also missed the leeches and ticks. In contrast to the rain, morning frost touched the open ground three times.

Our Australian experience almost ended in disaster. When returning our rental car to Europcar in Brisbane Airport, I left my wallet on the counter. The manager discovered it, came running after us, and caught us a mere couple of minutes before we reached the terminal building. Had we got there we would have been lost in the crowd, and I doubt if we would have discovered the loss before arriving in Vancouver. Despite business class comfort, we did not sleep a wink on our thirteen-hour return to Vancouver because the journey coincided with our internal daytime. On the face of it, we arrived on the same day as the departure and about four hours before we left.

25

Begin with an Operation,
End with a Medal

OUR FIFTY-FIFTH WEDDING ANNIVERSARY, in 2017, was unforgettable. I spent the morning in bed, feeling grim, perhaps affected by a fragment of infected Arctic Char on the previous evening or undigested roasted chestnut eaten just before I went to bed. I got up for a soup lunch. Then a prearranged phone call from the president of the University of Toronto gave Rosemary and me the astonishing news that we would receive honorary degrees on June 7. Apparently, I said all the right things, though without animation. I handed the receiver over to Rosemary to wipe with Purell before putting it down, and within one minute, literally, I vomited all the soup and more into the loo—and instantly felt much better.

The promised event began at the Mississauga campus, in the fiftieth-year celebration of its founding, where we engaged in conversation with undergraduates and graduates before we gave the Snider Lecture to a mixed audience. The last question we were asked after the lecture was, "In all those years on Daphne, how did you avoid conflicts and remain happy?" "Easy," was my reply: "Rosemary Is Always Right." I have said that before, and when challenged I have explained that even if not strictly correct it is close enough to be a workable formula. I have to smile to avoid my response being taken literally. A more serious answer is contained in the golden compliment we once received from Louis Lefebvre, who is an artist and writer as well as a biologist at McGill University. He told us that, in his mind, we epitomized equality, in our work and in our relationship, and that we reminded him of those ancient Egyptian carvings or paintings in which a man and a woman face each other and are depicted as being exactly the same height.

At the university's convocation in Toronto the next day, degree-day for Mississauga campus science students, we were given honorary Doctor of Science degrees. After delivering uplifting convocation speeches that drew upon our experiences in Canada, we sat next to the chancellor, Michael Wilson, and watched as students, two at a time, received their degrees and answered the question "What is next for you?" from this kindly, grandfatherly man. Answers: medical or graduate school for most, law for a few, several "don't knows," and one "I am going to follow my dreams." Many of these eager and enthusiastic students then shook our hands and congratulated us, even as we congratulated them. One was in tears before she reached the platform and shed many in her excitement on Rosemary's gown. I was impressed by the ethnic and national diversity of the graduating class.

After numerous photos outside with students and parents, we walked in lovely sunshine to Massey College, where a lunch was given in our honor. I sat next to the university president, Meric Gertler, and restrained myself from telling him I had vomited after our phone conversation. Rosemary and I had to give off-the-cuff responses to three well-delivered scripted speeches, mini eulogies. Neither of us is naturally good at this, although we may be getting better. I did get some laughs by recounting that at UBC my PhD hood had been put on by the president back to front (see Fig. 7.6), unlike today's experience. We talked to Peter Boag and Laurene Ratcliffe, who had driven over from Queen's University, in Kingston, Ontario. The reunion was a pleasant turning back of the clock to the beginning of the Daphne research, when their work helped to lay the foundation of the long-term study.

University of Toronto biology professor Marc Johnson was the brains and body behind the whole occasion, together with Spencer Barrett, who unfortunately could not be there. These two scientists are evolutionary biologists, and among other subjects they study the many ways in which plants and animals interact. We were very grateful to both, because it was a great honor to receive the degree from Canada's largest university and a splendid occasion in Convocation Hall in Canada's 150th year. Marc gave an extraordinarily powerful, moving, and eloquent address to the convocation in introducing us, before asking the chancellor to bestow the degrees.

Born on an island, Rosemary and I have a natural feeling for islands. In returning home from the Hanski symposium in Helsinki (chapter 23), we stopped in Iceland and stayed with Einar and Katrin Arnason. In the afternoon, Einar took us to see the house where Halldór Laxness lived for many years outside Reykjavík. Since he wrote about rustic life (Laxness 1999), we were surprised to see a modern-looking and substantial house complete with swimming pool in the garden, a small copse of birches, and a stream. It was the property of a fairly wealthy man and looked a lovely spot to reflect on life and write epic novels. I thought of Edvard Grieg's house at Bergen in Norway (chapter 5).

The next day we engaged about thirty students and postdocs in an open forum discussion of evolutionary biology and, after lunch, gave a seminar in the Nordic Center. The previous time (2009), Vigdís Finnbogadóttir, the former president of Iceland, sat in the front row of the audience and afterward asked us a good question about the seminar, but no ex-presidents attended this time.

We thought the country was wild after we left Reykjavík the next day and drove northwest, but it seemed tame when we returned the following day after traveling across much wilder terrain between a series of beautiful fjords. The first part of the journey covered some of the same ground that our Whitgift School expedition traveled in 1955. This time we drove through a long tunnel beneath the estuary just north of Reykjavík and passed a whaling station north of Akranes that I did not recognize, although it was probably one that I photographed in 1955. Some of the same farms were still there, I am sure, looking prosperous and with more buildings, yet they were still rare, human-imposed features on a wild landscape. Then we turned left into the western fjord region and finally reached Patreksfjördur.

The next day was our last full day, and we drove to the westernmost point of the island to marvel at the largest colony of seabirds in Europe on the sheer cliffs of Látrabjarg. The ledges were swarming with Brünnich's and Common Guillemots, including the bridled form. Razorbills were not much less abundant, and fulmars were nesting everwhere. At a distance on the rocks below sat two pure-white Ivory Gulls. Puffins guarded their nests on the clifftops, their incongruously colored beaks at odds with the landscape. On our way back over one of many moors between successive fjords, a mottled white ptarmigan flew up from in front of the car, and two golden plovers in gorgeous plumage tried to stare us down on the road. We stopped at a museum to look at a cornucopia of artifacts of a style of rural and marine life that was already a little old on my first visit and was now fast disappearing.

As in several past summers, we participated in the graduate course in Guarda, Switzerland. On the last day, I hurriedly put a lot of e-mail messages into both Rosemary's trash and mine. One of them came from "Awards"; clearly junk mail. Two weeks later, another came from Awards, and this time I read the subject: "RE Royal Society Royal Medal—Premier Awards." The message read: "I am trying to coordinate a date for this year's Premier Awards Dinner and I wanted to ask your availability for the following dates . . ." Only when the thank-you arrived in response to my reply did I think to scroll down for a meaning of this mystery and discover the original: "It gives me great pleasure to inform you both that you have been selected, subject to the Palace approval, as the winners of the Royal Medal B." Wow!

I turned to Wikipedia: "The Premier awards are the Royal Society's most prestigious medals and recognize exceptional and outstanding science. The Royal Medal of the Royal Society is a premier award. Known as the Queen's medals they are awarded annually by the Sovereign on the recommendation of the Council of the Society. They were founded by HM King George IV in 1825."

Life's pendulum then swung from the best to the worst. Settling back into life in Princeton, we learned from our friend Surin Vasdev that the small polyp he had removed from my colon a week earlier in Corvallis,

Oregon, was, surprisingly, cancerous. A couple of days after I had a CAT scan done, Toshio Nagamoto, a colleague of Surin, very kindly agreed to fit me into his busy schedule at very short notice and performed the necessary surgery. Fortunately, the pathologist's report gave no cause for further concern beyond several signs of age-related wear and tear. I stayed in the hospital for two nights, then transferred to a stylish rental property in town owned by Toshio's wife, Beverly, as I regained strength in comfort.

This was in October (2017), a crazy month, reminiscent of 2005 in that it began with an operation and ended with a medal. We traveled to London on my birthday to stay in the Royal Society premises for the Premier Awards dinner. It was a splendid, unforgettable occasion, preceded by a reception at which Prince Andrew—the Duke of York—presented the Royal Medals to us and to two others for their work in physics and applied science. We had been primed on how to address him before that: "Your Highness" first, then "Sir." Awaiting the medal presentation, I was explaining something about Darwin's finches to three or four people and had just said the finches had diversified rapidly in the past one to two million years, when someone said, "Where's that from?" I did not hear the question, apologized for my poor hearing, and asked him to repeat it, which he did. Even though I had bent closer, I still did not catch all the words, and another medalist, Melvyn Greaves, rephrased them for my benefit. So, I then described the essence of molecular dating, and the anonymous individual shot off a barrage of questions, including, "How do you know they didn't get to the islands on a boat?" Well, the questions were coming so thick and fast I hardly had the time to register the challenging idea of boat transport a million years ago. It was about this time, however, that I looked down at his lapel and realized he did not have a name tag, and thought, *Oh myyyyyy gosh, he must be the duke!* Minutes later, he gave a very good speech for five minutes before presenting the medals: one for us to share, three inches in diameter, gilt—silver electroplated with gold—and inscribed on the edge with our names. He then left.

The five-course meal was delicious. Each course was small enough to fit on the surface of our medal—I hardly exaggerate; the dishes were a triumph of quality over quantity. The meal was interrupted at one point for the presentation of the Royal Society's top award, the Copley Medal to Andrew Wiles, ex-Princeton, for the proof of Fermat's last theorem. I had a good conversation with John Skehel, a member of the Royal Society's council, and afterward a brief one with the president, Venki Ramakrishnan, and treasurer, Tony Cheetham, after everyone else had left. Why are Rosemary and I always the last to leave parties? This might be irrelevant, but it dawned on us we were all Americans!

Mel Greaves followed up our meeting at the Royal Society by inviting us to the Institute of Cancer Research in Chelsea and Sutton in London. He had written a stimulating book on cancer as a legacy of our evolutionary past (Greaves 2000); hence his interests intersected with ours. The center of lively conversations was evolution and what can be learned by treating cancer as an evolutionary deviation from normal cell function or as a community of competing and evolving species. It was fascinating. We were intrigued by the parallels with Darwin's finch hybridization, and by the possibility of genes being transferred between cells in different organs, in effect different ecosystems. If this happens, it could generate potent new gene combinations and metastasis of the cancer: for example, DNA containing a mutation originating in the stomach may circulate in the bloodstream in vesicles known as exomes and from there enter a cell in the kidney. We earned our visit by giving a seminar on Darwin's finches at the end of our three-day stay, concluding with a dinner at a lovely location by the Thames at Barnes.

In early February 2018 came the shocking news that we had been chosen to receive the BBVA Foundation Frontiers of Knowledge Award in Ecology and Conservation Biology. BBVA is the Banco Bilbao Vizcaya Argentaria. Why was it a shock? We knew we had been nominated for the award, and we knew we had to be near a telephone on February 5, but it was a shock because we had convinced ourselves we had no chance; in fact, we were laughing about it when walking home. Half an hour after the phone call, we had an interview by phone, from Spain. Two hours later we were being filmed and interviewed by a couple of cameramen from Bogotá. The fuss did not end there, as we dealt with urgent requests for pictures and videos of us working with finches. Congratulations poured in. We continued to send information to Madrid and gradually calmed down. I caught a cold.

An honored return to Madrid could not have been predicted on my first visit, in 1957, when I was so naive I had been tricked out of money not once but twice in one day (chapter 4). All that went through my mind in mid-June, when Rosemary and I were in Madrid to jointly receive the BBVA award. On Monday, the day after arrival, we were on duty all morning with press interviews, followed by a rehearsal of the ceremonies that lay ahead. The first ceremony was an evening gala concert at the Teatro Real, preceded by fanfare, a procession of the fourteen laureates from eight disciplines, and then short videos of each of us at work; our spring interview of hours was distilled into about one minute. The orchestra matched the occasion, playing Beethoven, Mahler, and an intriguing piece of modern music composed by one of the laureates, Kaija Saariaho. We thoroughly enjoyed it, but unfortunately she did not, because it was not performed in the way she would have directed if she had been allowed to. I suppose it was as if someone had given our own lecture on Darwin's finches in a different style and with quite different emphases.

The awards ceremony took place at the BBVA Foundation headquarters on Paseo de Recoletos the following evening; *recoletos*, I was told means "recluses." As instructed, we assembled in a long line while waiting for the minister of the environment, who was held up in traffic; eventually she got out of the car and walked. Our nicely organized line began to distintegrate while we waited, and Yale economist William Nordhaus captured the moment perfectly when he quipped, "We did not get our awards by being good at staying in line!"

In the ceremony itself, the newly appointed minister handed us our certificates and later gave a buoyant, upbeat speech on the need for conserving nature. Rosemary and I went to the podium, where Rosemary, in her Kyoto dress, delivered our joint acceptance speech in fitting style. After the last award was completed, we filed through the audience and out of the building, into the warm evening air to enjoy cocktails in the large entrance, screened from the traffic noise of the street by an arc of trees and shrubs. Manuel de Falla's musical piece *Nights in the Gardens of Spain* would not have been out of place. Rosemary walked into big compliments for her speech from my sister and brother-in-law, Sarah and Alex, and many others.

At 10:00 p.m. it began to get dark, and at 10:40 as the crowd began to thin out, we were whisked away to an adjacent building and into a live radio broadcast and interview. Hearing the translator was not easy, and on the spur of the moment I switched into impromptu Spanish and continued for the rest of the interview; simple and elementary my Spanish may have been, it was welcomed, despite the occasional linguistic glitch. We returned to the reception and eagerly sought another drink of red wine. We were in bed soon after midnight for the first deep sleep, finally having thrown off jet lag.

Having graduated with an unimpressive second-class degree from Cambridge in 1960, I could never have predicted a return fifty-eight years later to be installed as an honorary fellow of Selwyn College. I know what my chemistry tutor would have said! Rosemary and I socialized with my classmates Mike Young, Philip Crowe, and Chris Dobson at The Plough at Fen Ditton, and we all said, "You look just the same," which is code for "I can still recognize you." Later we met the college master, Roger Mosey, a friendly unpretentious man, ex-BBC. The installation took place in the chapel in the evening, and the ceremony was followed by dinner in the hall and concluded in the fellow's room—altogether delightful. Mike Young was the prime mover behind the scenes. He explained that the fellows had unsuccessfully tried to change the rules to allow Rosemary also to be admitted as an honorary fellow.

Three years before these awards in London, Madrid, and Cambridge, we visited South Korea for an award of a different kind. It was our second visit to South Korea, and the occasion was very special and unlike any other: the inauguration of Darwin's Way at the new National Institute of Ecology site near Gunsan. Jae Choe invited us for a few days, and the comfort of business-class travel made the invitation irresistible. On top of that was the lure of inaugurating a companion walk labeled Grant's Way (Fig. 25.1). This took place on November 23, 2015, close to the publication date of *Origin of Species* (chapter 20).

FIGURE 25.1. **Upper**: With Jae Choe on Grant's Way at the National Institute of Ecology at Gunsan, South Korea, November 2015. **Lower**: At a commemorative plaque on Grant's Way.

After speeches in front of about fifty people, much photography, and the unveiling of a plaque in front of a circular pool with islands representing the Galápagos, we set off on the inaugural walk. The path took us up through deciduous woodand and into pine forest covering the upper half of a small hill. At intervals there were slabs of polished stone with a couple of sentences inscribed in white letters on a black background describing in English and Korean significant events in Darwin's life: early days, voyage of the *Beagle*, and so on. Then the path forked, and Grant's Way began with the left-hand lower route.

The first marker of our research career along the path was an open hut representing the main cave on Daphne, complete with *chimbuzos* (water containers) and, miraculously, a member of our FIU (Finch Investigation Unit) on Daphne (chapter 12) in the early days: Stephen (Spike) Millington, an Englishman. His familiar broad smile separated an unfamiliar gray mustache from a grizzled beard. He was conveniently based in South Korea, working as a professional ornithologist in biological conservation. He joined us as we continued on the path downhill and past a few other markers on slabs of polished rock, just as on Darwin's Way. It was not a sand walk as at Darwin's home (Down House); rather, it was a muddy clay walk, which I thought was appropriate for us "muddy boots biologists," as a molecular biologist once disparagingly classified people like us. The path then rejoined the descending Darwin's Way, and the conjoined paths led to the final stone celebrating Darwin's achievements with a perceptively apt declaration composed by Jae: "Darwin's greatest contribution to humanity was to humble us, as a species, once and for all." Once and for all? I wish.

We gave a lecture afterward to a large and heterogeneous audience in a new, attractive auditorium, then visited the hands-on museum, which must be very stimulating for the children for whom it was designed.

Visiting two temples with our guides, Gilsang Jeong and Jongmin Kim, switched the mind to a different past. The first temple was Naesosa, originally built about fifteen hundred years ago. The site is memorable for a one-thousand-plus-year-old tree in the courtyard, a *Zelkova serrata*, massive in girth (approaching three feet), with several limbs shortened by half, which gave to the tree the impression of arthritic old age. The second temple, actually a cluster, was Seonunsa, in a slightly more remote and tranquil setting in pine-forested hills. Here we looked into the main Buddha temple and ancillary ones and helped ourselves to tea inside one devoted to hospitality.

At year's end we had the unexpected pleasure of news that our book *How and Why Species Multiply* (Grant and Grant 2008) was to be translated into Korean, thereby joining a Japanese and a Spanish edition.

Many years previously (2012), we had entertained Antonio Lazcano from Mexico City at home when he visited Princeton to give a lecture to the department on the origin of life. Our faculty colleague Laura Landweber had asked whether anyone would like to host him; we offered, and he was our guest for a whole week. Finally, in 2018, we managed to fit in a reciprocal visit.

Our lecture was at the Colegio Nacional in Mexico City. The *colegio* is in a converted nunnery that looks like the BBVA building in Madrid and is possibly a copy of it. The lecture itself was given to about 250 students and another 600 in remote, electronically connected locations. The only other member of this exclusive institution that we met was José "Pepe" Sarukhán, a very distinguished ecologist, ex-rector of the National Autonomous University of Mexico, and then in the government as, I think, subdirector for the environment. Rosemary enjoyed his company at the dinner afterward, while I enjoyed the conversation with his wife, Adelaide, and others within my limited hearing range.

We stayed with Antonio in his palatial apartment, richly adorned with books and art treasures. Breakfast was a two-hour conversation with our host, a learned, multilingual man who mixed gossip with erudite discussion on everything from ancient Mexico to ancient Rome. Thanks to his generous hospitality, we were able to spend hours on the first afternoon at the National Museum of Anthropology, hours on the second day at the Templo Mayor and its museum, and most of Sunday at the pyramids of Teotihuacán. This last extraordinary site was the exciting center point of our previous visit to Mexico City in 1962, at which time the palace of Quetzalpapálotl at Teotihuacán had not been excavated, and the Templo Mayor was underground and unknown.

Teotihuacán—which to the Aztecs was the place where gods were created—is in remarkably good condition for a city of twenty-five thousand people that was destroyed by fire fifteen hundred years ago. The
pyramids are points of contact with the spiritual world, as are the much
earlier ones in Egypt, the cathedrals of Europe, and the wall paintings
of Kakadu. The manner of making those contacts, however, differed
profoundly among cultures. The Mexican pyramids alone were associated with ritual sacrifice. We climbed both the Pyramid of the Sun and
the Pyramid of the Moon, without difficulty but with a horde of people.
At a distance, this must have looked like ants crawling over a pyramid
of sugar. We followed the ants into the well-preserved palace of Quetzalpapálotl to admire and wonder at the inspiration and meaning of
mythological birds and jaguars depicted in wall paintings. The museums
as well as the sites themselves are engrossing.

I have to add, the atmospheric pollution in Mexico City was awful,
rivaling that of Beijing. We were horrified to see it as we flew in, and
again as we descended from Teotíhuacan on our return into the main
city in what was once the basin of the old lake Texcoco.

We thanked Antonio one last time by e-mail on returning home, and,
true to his scholarly calling, his quick response was a quotation from
Cicero: *One of the true gifts of life is to have friends with whom one can talk
as with oneself.* Exactly!

26

Finch Genes and a
Return to Galápagos

I HAVE BEEN fascinated by what happened in the past ever since I became interested in my own history and attempted to reconstruct it with the fragments of memory. Questions arise from the large to the small, from the origin of life to the origin of Darwin's finches. What were the Galápagos like when the ancestral finches arrived? I would give anything for a time machine to help me find out! How did the radiation unfold and how was it guided or driven by changing circumstances? In the absence of fossils, we do our best to hazard a guess about history by studying the differences in the genomes of species and inferring the time course over which the differences evolved. Like my own history, the evidence is fragmentary.

We started by studying finches and finished by studying genes. Except that we haven't finished. The truth behind the statement is a collaboration with Leif Andersson at Uppsala University in Sweden to study genetic variation at the molecular level. Leif and colleagues had published a paper on the genetic analysis of skin-color variation, either white (or pink) or yellow, in chickens. We had wanted to study the genetic basis of a similar color polymorphism in the beaks of nestling finches, so we wrote to him in 2011 and asked whether we could work together, and he agreed. The result was identification of a single gene, BCO2, that determines whether carotenoids in the diet are deposited in beak tissues (yellow) or not (pink). This was the kind of connection between a gene and an avian trait that I had vaguely wondered about when I was an undergraduate and beginning to think of research (chapter 5).

From that simple beginning, an enormously rewarding collaboration prospered for a decade and continues. Leif and his group were able to reconstruct the probable evolutionary history of the finches by sequencing the genomes of all the species from the abundant samples of blood and DNA we obtained every year from 1988 onward. They made two important discoveries. First, they discovered that hybridization and gene exchange must have occurred early in the radiation. This was a gratifying extension of our findings of contemporary hybridization on Daphne, because it confirmed our conjecture that exchanging genes has been going on for hundreds of thousands of years between species and not just between populations of the same species. By the time of this discovery, the importance of hybridization in animal evolution was no longer the contentious subject it had been when we first studied it in the 1970s.

Second, Leif's group discovered two genes that influence body size and beak shape. HMGA2 and ALX1 are transcription factors that regulate the development of beak size and beak shape, respectively. Size and shape are the major features in which the finches have diversified, so these discoveries gave us powerful genetic tools to investigate evolution as a contemporary process. To do that meant applying those tools to the populations we had studied on Daphne. We first turned our attention to events in 2004–5, when *Geospiza fortis* with small beaks survived a drought better than those with large beaks. DNA was available for all banded *G. fortis* alive in 2004—seventy individuals, approximately half of which survived to 2005—so we compared the successes and failures and found a strong difference between them: only a third of the finches with the variant at the HMGA2 locus associated with large beaks survived, whereas almost two-thirds of those with the other variant associated with small size survived. Heterozygotes were intermediate in size and survival. This finding provided for the first time a demonstration of the evolutionary effects of natural selection at the level of individual genes affecting beak size (Fig. 26.1). Moreover, we knew how

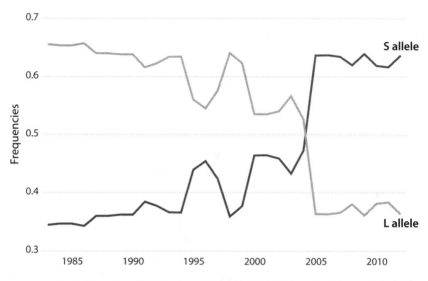

FIGURE 26.1. Change in allele frequencies at the HMGA2 genetic locus. S (small) and L (large) refer to their respective associations with beak size.

to interpret the effects: *G. fortis* with large beaks suffered in competition for large seeds from *G. magnirostris* (chapter 18).

From natural selection, we switched to the origin of the enigmatic Big Bird lineage (chapter 18). The outstanding question at the beginning was, Where did the original Big Bird (5110) come from? Convinced it was a Medium Ground Finch (*G. fortis*), we reasoned it must have come from one of the surrounding islands in line of sight from Daphne. Our analyses with microsatellite DNA from birds on only those islands suggested it was not a *G. fortis* but most likely a hybrid *G. fortis* × *G. scandens* that flew over from Santa Cruz Island. That is what we published, and we were dead wrong!

Leif and Sangeet Lamichhaney sequenced the genome of 5110, compared it with the genomes of all other species, and found that it matched a different species, *G. conirostris*, the large Española Cactus Finch! This species is found only on Española and its satellite island Gardner, more than sixty miles away on the southern periphery of the archipelago and well beyond the circle of islands around Daphne. It could have stopped at a couple of islands on the way, but still, Daphne is very far.

Remarkably, 5110 stayed to breed with *G. fortis* instead of turning around and flying back home in the wet season of the following year, as vagrants usually do. This discovery was both exciting and sobering. Exciting, because it revealed the Big Bird lineage was formed by two species hybridizing. We had written about this as a possibility after studying hybridization on Daphne for so many years, never expecting we would witness it. Simultaneously, the discovery was sobering because it made us realize how we went wrong. We should not have abandoned our separate and independent first impressions, inscribed in our notebooks (chapter 18), that 5110 was a *G. conirostris*, in preference for a more "plausible" alternative of *G. fortis* from a nearby island. It was a lesson in distrusting the law of parsimony and its beguiling simplicity.

One further insight into the genetic consequences of hybridization emerged from a close genetic analysis of hybridization of *G. fortis* and *G. scandens*. Although hybrids backcross bidirectionally and exchange autosomal genes, genomic analysis revealed almost no transmission of the Z sex chromosome from *G. fortis* to *G. scandens*. Our field observations provided the biological explanation: female (ZW)–biased gene flow, arising from hybrid-male (ZZ) disadvantage in competition for territories and mates, and no disadvantage to hybrid females.

By 2017 we knew further research progress was limited by the quality of the DNA, and we could learn much more about finch genomes if we could obtain high-quality DNA and run it through newly developed machines that read and record long sequences of the constituent nucleotides. Thus, six years after leaving Daphne, we returned to Galápagos for two weeks in March–April 2018. This time the destination was Puerto Baquerizo Moreno on San Cristóbal, and our companions were Leif and his wife, Barbro; Sangeet; and Carle "Calle" Rubin. Carlos Valle had worked wonders to deal with the permits and continued to do so up to and beyond our departure. The bureaucratic complexity of obtaining permission to do field research had increased markedly since 2012, and without Carlos's help we would have been stymied. Carlos aided in two other ways, first in providing us with the only lab facilities in Galápagos, at the University of San Francisco campus (Galápagos Institute for the Arts and Sciences, or GAIAS), and second in helping us with netting at sites near Socavón, Bosque (or Jardín) de Cactus, and Galáp-aguera, on San Cristóbal. This was a generous use of his time, because he was also giving a course on ecological field methods at GAIAS.

The trip was a big success. Over the two weeks, we caught all seven species of Darwin's finches on San Cristóbal Island and a total of forty-three finches in all. The Vegetarian Finch (*Platyspiza crassirostris*) was the most difficult, but playback of a previous tape recording drew one into the net, and a vicious mockingbird chase "helped" another. The most interesting birds were four *G. scandens*. As in past years, the measurements of a couple made us suspect they had hybridized with *G. fuliginosa*, which was confirmed by some students working in another group. We did not see or capture *G. magnirostris*. This species became extinct on San Cristóbal and Floreana Islands in the years after Darwin's visit in 1835, so rapidly that we are left to speculate on the causes: prob-ably a combination of killing by humans and destruction of *Opuntia* cactus and its seeds, an important food source. Reinvasion from another island always seemed possible, and in fact one had been seen and pho-tographed in recent years.

Offsetting to some extent the pleasure of capturing, handling, and measuring finches again were numerous *carmelito* (black-fly) bites and scratches from the introduced plant *Lantana camara*. Still, it was nice to see almost all of our familiar plants from Daphne and to walk in leafy and aromatic *Bursera* woodland again. In Calle's and San-geet's expert and dedicated hands, the nanopore-sequencing ma-chines they had brought worked very well with minute drops of fresh finch blood, generating vastly more information than our earlier as-says, with reads of sequences up to two orders of magnitude longer. The principal fruit of this labor was the discovery of twenty-eight regions in the genome associated with beak and body traits. That alone justified the trip. One day I hope we will be able to see whether the DNA of extant species contains signatures of extinct lineages or species, molecular ghosts of the past lurking in the genomes of the descendants, as has been achieved with other organisms, for example apes (Kuhlwilm et al. 2019).

The next year we were back again. At the beginning of July, the Galápagos National Park and the Charles Darwin Research Station jointly celebrated their sixtieth birthdays with a symposium in the lecture hall in the Miguel Cifuentes building on Santa Cruz Island. We were invited by Arturo Izurieta, the research station director, to give a keynote address, the only pure-science talk in a two-day symposium of applied science, parks management, and social studies. All the talks were in Spanish, yet even with hearing problems I did slightly better without the help of a translation. We prepared our talks in Spanish but changed our strategy on Arturo's advice, and instead I introduced our talk in Spanish, switched to English for the main part, and finished in Spanish. The lecture was well received, especially by the numerous guides, and we were photographed with clusters of happy young people countless times. There followed an afternoon of celebration by the park directorate alone, and then another evening at the station, where we met the British ambassador, a fairly young woman who adroitly handled our questions about Brexit. At these various times we met many old friends, including long-term residents Godfrey Merlen and Tui De Roy, and Simón Villamar, our guide on Fernandina, as well as making new ones, including Norman Wray, then president of the governing council of the Galápagos, and his wife, with the middle name Rosemary—improbable names for Ecuadorians. The mood was buoyant, celebratory.

Arturo kindly arranged for us to go on a day tour to Santa Fe Island and then to the Plaza Islands aboard the *Pikaia*, enabling us to see the islands as tourists see them. The islands were dry and looked the same as we had remembered them. More land iguanas were to be seen than finches. Arturo and his wife, Alexandra (Ale), also arranged for us to visit the Safari Camp in the western part of the Santa Cruz highlands to look at Daphne at sunset. This high-quality retreat, established in 2007, was new to us; we did not even know it existed. It is luxury in the highlands, right next to a national park forest in apparently excellent health.

Our week was very busy, yet we had two days off to socialize with Thalia, who, together with Devon, had preceded us by only a couple of days. Then on July 6 came the exciting highlight of our Galápagos visit, a one-day return to Daphne. Thalia joined us; so did Arturo and Ale, Juan, a photographer, and a new park ranger.

The sea was amazingly flat; I do not remember seeing it so calm, and we literally walked off the *fibra* (small boat) and onto the island. We did this in reverse on leaving, so we never set foot on the landing, the "welcome mat" (Weiner 1994), let alone face the challenges described in chapter 1. The goal of our visit was to search for Big Birds, with Thalia taking an independent route once we had climbed to the point above the craterlet and camp. Rosemary and I surveyed the craterlet and crater, while Thalia climbed out of the crater and on to the plateau.

The island had experienced a less-than-full blown El Niño event but was dry, and many finches were feeding quietly beneath cactus bushes, so it was difficult to see many of them clearly. Four were "our" banded birds; one of them still had all three color bands. With our observations combined, we saw approximately one hundred *G. fortis*, one hundred *G. scandens*, twenty *G. magnirostris*, and possibly up to six Big Birds. These were roughly the same proportions as on our last visit, in 2012 (chapter 21). The Big Birds gave us a problem because they did not look substantially larger than some deep-beaked *scandens*. We wondered whether Big Birds had bred with *scandens* and then backcrossed to them, and we started to think another trip to Daphne was needed. Song was almost absent; we heard just a few *scandens*, briefly. But surprisingly, two male *scandens* were feeding single juveniles, and so was a female *magnirostris*. Only two pairs of Blue-footed Boobies were breeding in the crater. For the first time in our experience, Magnificent Frigatebirds were nesting on the large patch of *Sesuvium* in the middle of the craterlet. I saw only one Greater Frigatebird.

As I stood in the crater and looked up, I could not help feeling amazed at what we had managed to achieve over many years in this physically demanding environment, when we had the energy to walk and climb all over it. I was seeing the island as others see it: topographically formidable and uninviting, the "ashcan" of the Galápagos, as someone from the *Reader's Digest* called it many years ago. Today our progress was slow and cautious, almost sedate, with frequent stops to look for birds. And then I disgraced myself by falling and scraping my arms and legs as

I attempted to stop a slide downhill. It happened as I turned to descend from a point on the northeast inner slope of the crater, when the rock I was standing on decided a better position was a yard farther downslope. Fortunately, I turned to fall into the slope, broke no bones, and twisted neither ankle. Remarkably, the glasses in my top pocket and my binoculars were not damaged. The worst effect was a stiff neck, possibly whiplash. I was seeing brown rocks as green, and then I started to lose focus, so I shut my eyes and rested on my back with my neck supported. Rosemary said I passed out briefly, but I was not aware of it. Everyone was anxious and solicitous.

After about half an hour of rest, I got up and was fine from then on— no mental stress, and everything in my visual world was correctly colored. I checked the fruiting of a couple of small cactus bushes that had become established on the crater floor a decade earlier. Then we climbed out of the crater, my wounds were treated with antibiotic ointment, and we had lunch. After lunch I enjoyed walking around the rim northward to the plateau point, where I paid my respects to the *Bursera* tree that had fallen before our first encounter in April 1973 but was still rooted and was looking as healthy as ever. It will surely outlive us. From there I continued along the rim and so on down to the path and farther to the landing. This was about the fourth time we left the island with sadness, for the last time, and this time I carried dermal souvenirs.

27

In Praise of Old Nassau

IN EARLY MARCH 2019 we gave a keynote address on speciation and hybridization at a Gordon Research Conference in Ventura, California. We had never attended one of these conferences, and I am glad we did, because it was one of the most enjoyable and rewarding meetings I have attended. The meeting of about seventy people, all at or above postdoctoral level, was designed and orchestrated by Rebecca Safran and Katie Peichel, experts in bird and fish evolution respectively. Size and focus were big factors in the success of the conference, and there were additional reasons: the standard of presentations was high, there was plenty of opportunity to socialize at mealtimes and at the posters, and the two hours or more after lunch were free time that we used to walk around the marina to a coffee shop for café latte. And the weather was sunny, with temperatures reaching 65°F.

On the morning of our departure for the Los Angeles airport, we switched on our computers to check e-mail and learned the shocking news that our long-time friend and Princeton colleague Henry Horn had died the previous day. We were not given any more information and were left to presume it had been a heart attack, as he died in the afternoon in his office, which had once belonged to John Bonner. John, a self-described maverick biologist (Bonner 2002) and an equally long-standing friend, had died a month earlier, soon after we had paid him the last of several visits at the ashram in Portland, Oregon. So, we lost two close friends in the space of two months. I was preoccupied with these sorrowful thoughts on much of the way to the airport, and then on the journey to Portland, to spend a few days with Nicola and Anjali (now self-renamed Rio). But I was able to appreciate the lovely Santa Monica hills, Mediterranean vegetation, and alluring hiking trails as we took the coastal route back to LA. We regretted not staying longer; we should have taken the opportunity of joining several colleagues on a one-day trip to nearby Santa Cruz Island. As we traveled around a bay to Malibu, I viewed some houses on the cliffside and had a sudden flashback to a hitchhiking ride I had with an Englishman and an American on this road in 1959. I had been on the way to Pasadena to stay with Walter and Mary Evans, whom I had met in Norway the year before. My long-term memory may be good, but the recall button is faulty, and this was a rare success.

June 4, 2019, was our day in the sun. Rosemary and I each received an honoray degree from Princeton University, in perfect, cloudless weather at 70°F. An honorary degree from your own institution is truly a great honor, given to few, the highest tribute you can receive from your colleagues, and a capstone to an academic career. President Chris Eisgruber gave an excellent short speech on civic values, and we listened to an impressive speech from the class of 2019 valedictorian, Kate Reed. The latter was admirably rich in text, meaning, and vocabulary, humble, unorthodox, and with a theme of uncertainty, vulnerability, and empathy. It was about openness to learn from others and humility. A couple of key sentences: "We [should] notice the starkly unequal distribution of kinds of vulnerability and voice. Princeton, in both its injustices and its ability to instill insight, has prepared us well to meet the world beyond the Fitz-Randolph gate with this kind of attention and awareness." It was courageous to use the word "injustice," and spoken mainly for minorities and women in particular, as many students were then protesting at the university's perceived failure to apply the federal government's Title IX provisions against sexual harassment and abuse. In fact, at the conclusion of the degree ceremony, when everyone was singing the Princeton signature tune, "In Praise of Old Nassau," a group of about thirty or forty graduates at the back of the audience turned away from the podium in front of Nassau Hall and faced Nassau Street as a gesture of protest.

I noticed their protest only at the end; for most of the time I was transfixed by the audience's performance of a strange, coordinated gesture, truly and traditionally Princetonian. Toward the end of the song, each person clenches the right hand close to the left shoulder, and then emphatically and in unison lets the retaining muscles go and extends the arm, almost throwing it outward and bending it in a dangerous arc toward the neighbor on the right, then reversing it, then repeating it, all in unison and coordinated with the words of the song. I was mesmerized by this synchronized behavior because it is reminiscent of the displays of fiddler crabs on sandy beaches or muddy shores (Crane 2016). I suppose left-handed people switch to being right-handed to avoid a fight.

The procession left the podium, and we walked past the audience of students, parents, and friends, an uplifting sight of multicolored robes. One could scarcely wish for a more compelling image of ethnic and racial harmony in these troubled times, and it was very nice to see the happy, laughing students. The graduating class included a diversity of low-income and first-generation college students the university recruits—something it does right. Afterward we had lunch in the Chancellor Green rotunda, which is the original university library, and chatted with Edith Grossman, a fellow honorary degree recipient and a famous translator of Spanish-language works (Miguel de Cervantes, Gabriel García Márquez, Mario Vargas Llosa); Scott Berg, a writer and biographer; Ellen Ochoa (another degree recipient), the first Hispanic woman in space; and Anne Sherrerd, a member of the board of trustees (like Scott) and our "minder" or escort.

The buildup to receiving the honorary degrees actually started the evening before with the president's dinner in Prospect. I sat next to Kathryn (Katie) Hall, the chair of the board of trustees, and had a fine, far-ranging conversation. I was impressed by her and by other members of the board we met then and on the following day; they were much less the stern, investment-manager types that I had envisaged and much closer to friendly academic types. At the dessert, one by one, we degree recipients were invited to a lectern to deliver a few words on what the degree meant to us personally.

When Chris introduced us, he mentioned that the board of trustees had broken its rule in awarding degrees to a husband-and-wife team for the first time. Rosemary's response was short, compact, warm, and effective, ending with the hope that the rule will be broken many times in the future. Mine was a bit more spontaneous, rambling about how unpredictable my life had been, starting in prewar London, surviving the bombing, and later emigrating to North America; I finished with a warm, from-the-heart, thank-you and sat down. Afterward, I rearranged the words and gave a much better, more fluent talk in my head, so it came as a surprise when about a half dozen people came up to me and thanked me for a very moving speech.

The speech given by Ellen Ochoa was much more polished, and dramatically effective when she told us the space shuttle she had been so much involved in would be passing over Princeton at nine o'clock. And sure enough, there it was when we walked out to watch, bright in the cold, clear night sky, moving toward our home and beyond. I am sure it flew over our garden, a new bird for the garden list.

Our peripatetic lifestyle resumed with a visit to Costa Rica. It was a first visit to that country for us, apart from the trip to Cocos Island in 1997. The purpose was to give a keynote address at the Neotropical Ornithological Congress in San José. Unfortunately, we could visit for only two days; however, the conference was shortened by one day, allowing us time for exploration. After the lecture we were photographed countless times with countless young people from Argentina, Brazil, Peru, Chile, Ecuador, Costa Rica, Panama, Guatemala, and Uruguay. They thanked us for inspiring them, very movingly—as educators we can wish for no higher praise. We left the intoxicating Latin rhythms early, at nine p.m., to get some sleep because we had to get up at four a.m. for a bird-watching trip to the highlands with biologist Diego Ocampo and an experienced ecotourism guide, Esteban Biamonte. After a two-hour drive we reached the Paraíso Quetzal Lodge in the Savegre–Cerro de la Muerte region at seventy-two hundred feet elevation. On the way we passed hundreds of pilgrims in small groups walking to a shrine that commemorates a miraculous visitation of the Virgin Mary, which an indigenous woman had witnessed four times. Some carried small cloth bundles, others very little. Perhaps hospitality was offered en route.

Before stopping at our hotel destination for breakfast and a couple of hours of bird-watching, we hit the highlight of the day, a magnificent view of a Resplendent Quetzal. It was a brightly colored male with a very long tail that undulated with the rhythm of a snake as it flew. The bird-snake in flight gave us an unexpected insight into the origin of the mythological Quetzalcoatl (plumed serpent). We first saw the bird feeding on fruits of a wild avocado—remarkably there are sixty species of avocado—then in picture-portrait position high up in the same tree, sitting quietly and with head still but yellow eyes permanently scanning the vegetation for insects. It flew one hundred feet in full view, allowing us to see the crimson chest and belly in contrast to the iridescent green of the head and throat and the undulating tail. No more than 150 feet separated us from this exquisitely colored bird. The reason we were so fortunate with the quetzal is that the local farms support ecotourism by providing information to a communication network on birds currently on each of the properties, thus we had a tip-off. Jorge, son of the hotel owner and our local contact, received the tip and guided us skillfully and informatively.

Many of the species in the vicinity were endemic, including brilliant hummingbirds at feeders, cryptic wrens in tangled vines, and a briefly viewed Spotted Wood Quail. From the hotel we drove up to the páramo, at 11,500 feet, which is not open country but shrubby and dense with thickets. In the misty environment, all we saw were two yellow-eyed Volcano Juncos, clearly and about thirty feet away, feeding rapidly on small insects. Surprisingly, they had colored bands on their legs. The plants were interesting. One flowering *Hypericum* looked like the Galápagos *Darwiniothamnus*, which grows in similar habitat. As we drove down, the rain soon began, and as it intensified on the way back to San José, I started to catch up on lost sleep.

A trip to Hungary in November 2019 was destined to be our last international trip for more than two years, for unexpected reasons described in the next chapter. Our hosts in Budapest, two theoretical ecologists, Liz Pásztor and Géza Meszéna, booked us into the Hungarian Academy of Sciences apartments. Our palatial room, with furniture of the eighteenth century, looked as if it had been prepared for an evening of music from a Mozart string quartet. We were not; we promptly fell asleep on arrival.

The next morning, after a long discussion on Darwin's finches, niche theory, and evolution, we traveled by train to Debrecen, in the eastern part of the country. Tamás Székely, a behavioral biologist from the University of Bath, in England, took us to his farm at Hortobágy the next day, where we admired his dark brown sheep—together they stand, together they fall, permanently touching each other—and his puli, a massive local sheep dog with matted hair too thick to see through. I had hoped to see a bustard but had to settle for a buzzard. However, we did see some cranes flying in, a minute fraction of the one hundred and fifty thousand that normally migrate there. A herd of Przewalski's Horses grazed in the distance.

For the next two days we participated in a symposium on reproductive strategies organized by Tamás and held at the University of Debrecen. Twenty-eight years beforehand, we had been in the same academy apartments and the same university for the European Society for Evolutionary Biology meeting. One experience on this second visit was a complete surprise. We discovered a small zoology museum named in honor of a distinguished alumnus, Miklos Udvardy, none other than my short-term supervisor almost sixty years ago.

We also made two seminar visits within the United States in October, first to Montana for four days and then one day at Stanford, California, to give the second Ehrlich-Mooney Lecture in Ecology and Evolution. Visiting other biology departments is fun when the agenda is full of interactive meetings with graduates, postdocs, and faculty in small groups. At Montana we gave two lectures. The students seemed to get as much pleasure from my part of the unrehearsed second one on mate choice, as I stumbled and tripped over my tongue, because, so they told me, it gave them encouragement. The Stanford visit in beautiful weather was very different, as we had meetings with faculty during the day, met graduates at the standard pizza lunch, and attended a small reception after our lecture. At the end of the evening dinner, we were given a witty send-off by Hal Mooney, a distinguished plant ecologist and part of the reason for our visit: "When you have got your research together, you should come back and tell us about it!"

28

And So to Lockdown

NOTHING PREPARED US or anyone else for the paralyzing and murderous events of 2020 that closed down the travel and lifestyle I have described in the last few chapters.

The tranquil beginning of the new year in Princeton was a cloudy day with a little sun and gusts of wind. We walked on the far side of Carnegie Lake to Kingston and back along the road, seeing fifteen species of birds, including many Common Mergansers and two immature Eastern Bluebirds. We celebrated the new year with mulled wine, mince pies, and Stilton cheese by a blazing fire, listening to Christmas carols. It was the first time we had celebrated like this since 2014. We needed a brief walk in the late afternoon and then completed the day by the fire with food and drink in moderation. I read my diary for 2019, jogging the memory of salient events and in their proper sequence. This was followed by our fifty-eighth wedding anniversary celebration in similar style. In the morning we walked by the canal at Griggstown in fog. The evening was warm and musical by the fire, sharing reflections, glad to have married in 1962 rather than 2020.

In early February we flew to Oregon to spend several days with Nicola in Corvallis. We gave her the opportunity to fly to Tampa, Florida, to take a short course in medical writing, and she came back thrilled with the experience, so we felt rewarded. While she was away, old man Blaze, the family's Australian shepherd dog, took us for daily walks, less ambitiously than in the previous year. When the weather changed for the better, we took stronger exercise by climbing to the benches at a lookout on a nearby oak-fir forested hill of the McDonald Properties.

Leaving Corvallis, we traveled to Texas A&M University at College Station to give a talk, in celebration of Darwin's birthday, on February 12. Leif Andersson, who has an appointment at the university, was our host on the greatly spread-out campus at College Station. The Darwin Day event was massive, with an audience of about five hundred. Our lecture was preceded by a community event with many students at booths and young children and parents enjoying the occasion to the full. Two skunks in a small (indoor) enclosure were the most popular exhibit. On the weekend, we joined Leif and two dozen keen birders on the Gulf coast in the vicinity of Corpus Christi. This was dedicated birding. One of our companions saw 151 species in total, and our tally could not have been very much less—more than all the species we saw in Australia in one month! The sightings included some Mexican exotics, such as a Green Jay and a Clay-colored Thrush, and a couple of Greater Roadrunners. We had not seen roadrunners for decades. Hawks there were aplenty. Among the plethora of large and small, colored and plain, the highlight was seeing four Whooping Cranes in Goose Island State Park, which is the site of a thousand-year-old live oak known as the Big Tree. These rare birds looked regally graceful and elegant as they came floating into land to join about eight others. Sandhill Cranes looked ordinary by comparison, and Roseate Spoonbills that should have stolen the show were eclipsed.

And then came the blow that struck us all, approaching by stealth and setting the world on a new course. While we were in Texas, if not earlier, the SARS coronavirus-2 that had caused Wuhan in China to "shut down" began to spread around the world, originating, apparently, from a transfer from bats or pangolins to humans. Mortality rates varied among countries from 1 to 10 percent, with first Iran and then Italy reporting shocking increases. In the United States, test kits, surgical masks, and other hospital equipment were in woefully short supply before any cases were announced, and hospitals were predicted to be overwhelmed later in the disease's progress. Despite that, at one point, President Donald

Trump said the virus would go away in a few days. Nonetheless, a few days later, an emergency fund of billions of dollars was set up. Panic buying and hoarding ensued. Toilet paper and hand sanitizer were the first to run out, followed by frozen foods. We were assured that the restrictions would continue for another month at least. The more cautious Canadian estimate was three months.

The initial political response to a crisis in either the natural or social environment is denial. Putin declared the virus was not in Russia but gave everybody a week's holiday anyway. Alexander Lukashenko, the president of Belarus, denied the virus existed in his country because he could not see it flying around. Instead, he said, coronavirus is a psychosis that can be fought with vodka, saunas, and driving tractors! Jair Bolsonaro, president of Brazil, more concerned with the economy than lives, said, you don't close a car factory after a car accident. True. You don't close a hospital after someone dies either, but how helpful is that? For truth-seeking scientists, such attitudes are literally incredible. On the subject of make-believe, gun shops in the United States were declared an essential service and so kept open, and long lines of buyers formed. Why are guns essential in a disease-caused crisis? Because, so we read, food will become scarce, and people will have to defend themselves and their food. Against whom? Well, against other people defending themselves and their food . . . and so on. A federal judge ruled that the state of Massachusetts violated the rights of citizens to defend themselves when it closed gun shops, and ordered them reopened.

It turned out later that the virus must have arrived in the United States in January, and not in February as previously believed, because someone in California had died with it on February 6. We stockpiled food and paper goods as best as we could. By the end of March, numbers of cases of infection and deaths were rising exponentially, and the United States overtook the rest of the world. The pharmaceutical industry launched a massive research program into the development of a vaccine, with extraordinary success.

I started rereading The Plague (Camus 1960) after an interval of fifty or sixty years. Although allegorical and not written to forecast the future, it described a progression of reactions and behavior remarkably similar to the responses in the current pandemic: initial disbelief, excuses that it will not last long, and so forth. Here is one small example: In early 2020, the World Health Organization (WHO) took the position that face masks were needed in hospitals but were not recommended elsewhere, because they had not been demonstrated to be effective and might give rise to a mistaken belief in protection (the policy was later reversed). In *The Plague,* the journalist Raymond Rambert asks Jean Tarrou if cotton-wool masks enclosed in muslin are of any use, and Tarrou replies, no, but they inspire confidence in others (Camus 1960, 192).

Having finished this marvelous novel, I turned to Defoe's *A Journal of the Plague Year.* It is similar to Camus's book in its descriptions of the people's reactions to a mysterious, bewildering, and threatening event, as told by a narrator who is both observer and interpreter; fear and disbelief, the spread of misinformation, and protests at being forced to stay in locked houses all have echoes in contemporary society. Northwood (Norwood), my place of origin (chapter 2), was a haven for those who fled from London. I read the book not for reasons of morbid curiosity but through a fascination with the similarities and differences between reactions to "an unknown distemper," then and now. Defoe describes people behaving as if the disease were caused by a coronavirus transmitted by inhalation or direct contact in some way. For example, merchants who brought supplies to Britain by boat were paid with coins that had been dipped in vinegar and thus "sanitized." In fact, the plague bacteria were transmitted by fleas, although there may have been some additional transmission from contaminated goods and clothing (Barbieri 2021).

To return to the beginning, on March 11, the WHO declared the outbreak to be a pandemic. Daily bulletins from the BBC on our computer screens continued to hypnotize us with numbers of afflicted people, numbing the brain. And then came the official beginning of spring, March 21, and a cause for personal celebration, with the news that one of our papers was about to be published and another was accepted for publication after an eleven-month wait. We celebrated by the fire with a salmon and Beaune supper.

At nine p.m. the telephone rang with a message: the governor of New Jersey had ordered a curfew, and everyone had to stay at home except for visits to food stores or pharmacies and once-a-day exercise—for the foreseeable future. All our planned trips, to São Paulo (Brazil), Virginia, Detroit, Berkeley, Oregon, and Vancouver, had to be canceled. The prospect of an alleviating vaccine was beyond the horizon, some eighteen months away at least. How devastating this would have been if it had occurred during our long-term study of Darwin's finches and prevented us from visiting the Galápagos. How disruptive it must be for scientists who currently have long-term research projects in other countries; members of our own department conduct research in Kenya, Peru, and Panama. We could identify with them and felt intensely sympathetic. Suddenly our precarious dependence on the natural world being treated with respect was forcefully brought into sharp focus by a threat to human lives—a self-inflicted wound through mismanagement of our environment. We were in lockdown, isolated as on an island like Daphne but without the enchantment. We went to bed talking of crises and World War II.

Six months later I started writing.

Epilogue

A SECONDARY THEME runs through this book and occasionally breaks the surface like the dolphins in the Tiputini River. It is the smile from Lady Luck. It has meant that important decisions went one way and not the other. I had the good fortune not to be blown to pieces in World War II, when more than seventy-five hundred children in London suffered that fate. I escaped catching tuberculosis, amoebic dysentery, polio, and other potentially fatal common childhood diseases in the first five years of my life. I had the good fortune to receive a fine education, narrowly qualifying for entrance to Cambridge University. And I had the immensely good fortune to meet Rosemary. At one point the odds were fifty-fifty of this happening, a spin of the coin. Coincidentally, before we ever met, we applied from Britain for graduate work (me) and a research position (Rosemary) at the University of British Columbia and the University of California at Berkeley. Berkeley said no, to both of us, maybe next year. Berkeley could have said yes to one of us after UBC had said no. Thank you, UCB and UBC. The list of good fortune is a long one. Even now I can scarcely believe it.

ACKNOWLEDGMENTS

IN WRITING this autobiography, I have been helped by wise advice in two books. *A Swim in a Pond in the Rain* by George Saunders (2021) provides excellent advice on composition and structure and showed me how Russian authors of short novels achieved their effects in holding the reader's attention. Thank you, Sean McMahon, for suggesting it.

William Zinsser's *Writing about Your Life* (2004) was more directly relevant to my writing, often arguing by example. He emphasizes being clear on who the intended audience is while avoiding attempts to guess what the audience wants the author to write. "Write what you want to write" is the essence of his advice, while always remembering there is an audience who wants to know what is meaningful to you. Thank you, Jonathan Weiner, for pointing me in the Zinsser direction.

In thinking about how to weave together personal stories and professional experiences, I benefited from reading autobiographical books by six biologists: Charles Darwin (Barlow 1958), John Bonner (2002), Evelyn Hutchinson (1979), Ed Wilson (1994), Ilkka Hanski (2016), and Jim Murray (2019), excellent storytellers, all of them.

A first version of the text was read by my sister Sarah Marianos and two close friends, Mark Chapin and Uli Reyer. They represent the three important components of my intended audience: a nonscientist and educator, an intelligent layperson, and a professional biologist with interests similar to mine. Their comments, advice, and suggestions are too numerous to mention individually, but collectively they helped me to change the structure and achieve greater clarity.

My cousin Brian Jackman read and helped with some details of our shared childhood. Ian Abbott read chapters 2 and 10. Daughters Nicola and Thalia made a few influential suggestions on how best to tell the story. Two anonymous reviewers forced me to rethink a large number of issues concerning overall design as well as detail.

My wife Rosemary read the entire manuscript more than once, corrected errors, and provided a series of insights that had escaped me. Discussing her nascent autobiography (B. R. Grant MS) helped each of us to develop our own.

To Alison Kalett at Princeton University Press, I am indebted for excellent editorial advice on how to achieve a balance between scientific exposition and description of everything else. Amy K. Hughes gave me outstanding help in copy-editing the manuscript and correcting numerous errors in names and places. I thank Dimitri Karetnikov once more for excellent help with illustrations.

Darwin's Finches

Charles Darwin knew a thing or two, yet a couple of others
   escaped him
With laser-like insight before lasers were invented
He knew the principal of common descent
and how the best were selected

He missed a piece in the mystery of life, and it was really
   very simple
Just a four-letter word beginning with G
Conjuring the magic of inheritance
and ending with the letter E

If the gene is a bottle, the elixir of life, there must be something
   inside
A genie if you like with a language of its own
And instructions on how to proceed
To ensure its seeds are sown

We've improved on that notion by releasing the genie without
   discarding the gene
By opening the genome to see what's inside
And finding a jumble of letters
That act as a coded guide

And here's where we pick up the story of finches whose name
   is linked to the man
He found them on the Galápagos Islands,

Among the slag of volcanic cones
In wandering little bands

The evolution of these drab brown birds and their adaptive
     radiation,
Was not in the realm of his comprehension
Though he noticed their different beaks
But then lost concentration

Some of these otherwise ordinary birds drink the blood of
     other birds
Another type uses a twig as a tool
If it cannot find a cactus spine
To use in a manner cruel

We have tried to discover how they evolved, and the role of
     natural selection
As well as the way in which they form a pair
According to beaks and the song they sing
And their willingness to share

In doing all this we found that they breed sometimes with
     another species
They exchange their genes, a providential gift
That helps their offspring attract a mate
And gives survival a lift

The biggest surprise on Daphne Major was the beginning of a
     brand new species
Forged from a union that determined its fate
Two species giving rise to one
That failed to reciprocate

We call them the Big Birds for that's what they are, they are
     breeding all by themselves
By combining the niches of three other species
With future uncertain but possibly long
If they avoid diseases

The Galápagos Islands were once called enchanted yet failed to
    enchant our hero
They've enchanted the finches and enchanted us too
We wonder if they are only a dream
Are the Big Birds really true?

REFERENCES

Page locations for works mentioned in this book are listed in brackets at the end of each reference listing.

Anderson, W., and C. Hicks. 1978. *Cathedrals in Britain and Ireland*. Charles Scribner's Sons, New York. [page 46]

Barbieri, R. 2021. Origin, transmission, and evolution of plague over 400 y in Europe. *Proceedings of the National Academy of Sciences USA* 118: e2114241118. [page 304]

Barlow, N., ed. 1958. *The Autobiography of Charles Darwin, 1809–1882. With original omissions restored. Edited with Appendix and Notes by his grand-daughter*. Collins, London. [pages 38, 112]

Beebe, W. 1924. *Galápagos: World's End*. Putnam's, New York. [page 119, 170]

Bonner, J. T. 2002. *Lives of a Biologist: Adventures in a Century of Extraordinary Science*. Harvard University Press, Cambridge, MA. [page 135, 294]

Bonnet, T., et al. 2022. Genetic variance in fitness indicates rapid contemporary evolution in wild animals. *Science* 376: 1012–16. [page 227]

Brodie, E. D. III. 2011. Natural history first (but don't stop there). *Evolution* 65: 3336–37. [page 209]

Bronowski, J. 1959. *Science and Human Values*. Harper Torchbook, New York. [page 112]

Browne, J. 1995. *Charles Darwin: Voyaging*. Vol. 1. Alfred A. Knopf, New York. [pages 110, 176]

Calsbeek, R. 2011. Cracking the seeds of evolutionary causality. *BioScience* 61: 828–29. [page 234]

Camus, A. 1960. *The Plague*. Penguin Books, Harmondsworth, Middlesex, UK. [pages 303, 304]

Carson, Rachel. 1962. *Silent Spring*. Houghton Mifflin, Boston. [page 87]

Clottes, J. 2016. *What Is Paleolithic Art?* University of Chicago Press, Chicago. [page 269]

Connell, J. 1961. The influence of interspecific competition and other factors on the distribution of the barnacle *Chthalamus stellatus*. *Ecology* 42: 710–23. [page 98]

Cox, P., ed. 2013. *Memories of Whitgift: The Boys' Own Tales, 1880–1980*. Labatie Books, Cambridge Book Group, Cambridge, UK. [pages 21, 29]

Crane, J. 2016. *Fiddler Crabs of the World*. Princeton University Press, Princeton, NJ. [page 296]

Dampier, W. 1927 (1697). *A New Voyage Round the World*. Argonaut Press, London. [pages 120, 151, 228]

Darwin, C. 1839. *Journal of Researches into the Geology and Natural History of the Various Countries visited by H.M.S. Beagle*. Henry Colburn, London. [page 111]

Darwin, C. 1859. *On the Origin of Species by Means of Natural Selection*. J. Murray, London. [pages 210, 265]

Defoe, D. 1966 (1722). *A Journal of the Plague Year*. Penguin Books, Harmondsworth, Middlesex, UK. [page 304]

Dobzhansky, T. 1937. *Genetics and the Origin of Species*. Columbia University Press, New York. [page 166]

Durrell, G. 1956. *My Family and Other Animals*. Rupert Hart-Davis, London. [page 107]

Edmondson, W. T. 1991. *The Uses of Ecology. Lake Washington and Beyond*. University of Washington Press, Seattle. [page 227]

Eiseley, L. 1970. Introduction. In *Galápagos: The Flow of Wildness*, vol. 2, *Discovery*, pp. 22–38. Sierra Club, San Francisco, CA, and Ballantine Books, New York.

Emlen, J. M. 1974. *Ecology: An Evolutionary Approach*. Addison-Wesley, Reading, MA. [page 100]

Fritts, T. H., and P. R. Fritts. 1982. *Race with Extinction: Herpetological Field Notes of J. R. Slevin's Journey to the Galápagos, 1905–1906*. Herpetological Monograph No. 1. Herpetologists' League, Lawrence, KS. [page 1]

Gesell, A., F. L. Ilig, and L. B. Ames. 1974. *Infant and Child in the Culture of Today: The Guidance of Development in Home and Nursery School*. Harper and Row, New York. [page 94]

Grant, B. R. MS. One step sideways, three steps forward. [pages 65, 89, 94, 97, 103, 107, 118, 132, 155, 167, 192]

Grant, B. R., and P. R. Grant. 1989. *Evolutionary Dynamics of a Natural Population: The Large Cactus Finch of the Galápagos*. University of Chicago Press, Chicago. [page 144]

Grant, P. R. 1972. Interspecific competition among rodents. *Annual Reviews of Ecology and Systematics* 3: 79–106. [page 99]

Grant, P. R. 1986. *Ecology and Evolution of Darwin's Finches*. Princeton University Press, Princeton, NJ. [pages 109, 143, 179]

Grant, P. R. 2000. What does it mean to be a naturalist at the end of the twentieth century? *American Naturalist* 155: 1–12. [pages 169, 241]

Grant, P. R., and B. R. Grant. 1997. The rarest of Darwin's Finches. *Conservation Biology* 11: 119–26. [page 174]

Grant, P. R., and B. R. Grant. 2008. *How and Why Species Multiply: The Radiation of Darwin's Finches*. Princeton University Press, Princeton, NJ. [pages 176, 182, 284]

Grant, P. R., and B. R. Grant. 2014. *40 Years of Evolution: Darwin's Finches on Daphne Major Island*. Princeton University Press, Princeton, NJ. [pages 2, 109, 150, 156, 191, 193]

Greaves, M. 2000. *Cancer: The Evolutionary Legacy*. Oxford University Press, Oxford, UK. [page 279]

Greenblatt, S. 2011. *The Swerve: How the World Became Modern*. W. W. Norton, New York. [page 241]

Haldane, J. B. S. 1932. *The Causes of Evolution*. Cornell University Press, Ithaca, NY. [page 241]

Hanski, I. 2016. *Message from Islands: A Global Biodiversity Tour*. University of Chicago Press, Chicago. [pages 38, 260]

Harrer, H. 1953. *Seven Years in Tibet*. Rupert Hart-Davis, London. [page 186]

Hasty, W. 2011. Piracy and the production of knowledge in the travels of William Dampier, c. 1679–1688. *Journal of Historical Geography* 37: 40–54.

Hutchinson, G. E. 1979. *The Kindly Fruits of the Earth: Recollections of an Embryo Ecologist*. Yale University Press, New Haven, CT. [pages 28, 82, 83, 244]

Huxley, L. 1900. *Life and Letters of Thomas Henry Huxley*. Vols. 1 and 2. Macmillan, London. [Pages 8, 138]

Isherwood, C. 1939. *Goodbye to Berlin*. Penguin Books, Harmondsworth, Middlesex, UK.

Jackman, B. 2021. *West with the Light*. Bradt, London. [pages 7, 18, 138]

James, C. 1980. *Unreliable Memoirs*. Jonathan Cape, London.

Jirinec, V., et al. 2021. Morphological consequences of climate change for resident birds in intact Amazonian rainforest. *Science Advances* 7: 1743. [page 236]

Kaplan, A. 2012. *Dreaming in French*. University of Chicago Press, Chicago.

Kazantzakis, N. 1952. *Zorba the Greek*. Transl. Carl Wildman. John Lehmann, London. [page 104]

Keynes, R. 2002. *Fossils, Finches, and Fuegans*. HarperCollins, London. [page 111]

Kuhlwilm, M. S., S. Han, V. C. Sousa, L. Excoffier and T. Marques-Bonet. 2019. Ancient admixture from an extinct ape lineage into bonobos. *Nature, Ecology and Evolution* 3: 957–65. [page 290]

Lack, D. 1947. *Darwin's Finches*. Cambridge University Press, Cambridge, UK. [pages 64, 109, 112, 152, 166]

Lamichhaney, S., et al. 2015. Evolution of Darwin's finches and their beaks revealed by genome sequencing. *Nature* 518: 371–375. [page 112]

Laxness, H. 1999 *Independent People*. Trans. J. A. Thompson. Harvill Press, London. [page 276]

MacArthur, R. H. 1972. *Geographical Ecology*. Princeton University Press, Princeton, NJ. [page 77]

Mayr, E. 1963. *Animal Species and Evolution*. Belknap Press, Cambridge, MA. [page 152]

Medawar, P. 1984. *The Limits of Science*. Oxford University Press, Oxford, UK. [page 191]

Murray, J. D. 2019. *My Gift of Polio: An Unexpected Life from Scotland's Rustic Hills to Oxford's Hallowed Halls and Beyond*. Chauntecleer.com. [page 138]

Nelson, J. B. 1968. *Galapagos: Islands of Birds*. William Morrow, New York. [page 120]

O'Brien, C. C. 1970. *Camus*. Fontana Books, London.

Oster, G. F., and E. O. Wilson. 1979. A critique of optimization theory in evolutionary biology. In *Caste and Ecology in the Social Insects*, chap. 8. Princeton Monographs in Population Biology 12. Princeton University Press, Princeton, NJ. [page 231]

Pant, S. R., M. A. Versteegh, M. Hammers, T. Burke, H. L. Dugdale, D. S. Richardson, and J. Komdeur. 2022. The contribution of extra-pair paternity to the variation in lifetime and age-specific male reproductive success in a socially monogamous species. *Evolution* 76: 915–30. [page 227]

Percy, F. G. H. 1991. *Whitgift School: A History*. Whitgift Foundation, Croydon, UK. [page 21]

Peterson, R. T., and J. Fisher. 1955. *Wild America*. Houghton Mifflin, Boston. [page 54]

Pianka, E. R. 1974. *Evolutionary Ecology*. Harper and Row, New York. [page 100]

Price, T. 2008. *Speciation in Birds*. Roberts and Co., Greenwood Village, CO. [page 265]

Reilly, J. 2018. *The Ascent of Birds. How Modern Science Is Revealing Their History*. Pelagic Publishing, Exeter, UK. [page 236]

Saunders, G. 2021. *A Swim in a Pond in the Rain*. Random House, New York.

Sacks, O. 2002. *Uncle Tungsten. Memories of a Chemical Boyhood*. Vintage Books, New York. [page 7]

Sankararaman, S., S. Mallick, N. Patterson, and D. Reich. 2016. The combined landscape of Denisovan and Neanderthal ancestry in present-day humans. *Current Biology* 26: 1241–47. [page 152]

Schacter, D. L. 2001. *The Seven Sins of Memory: How the Mind Forgets and Remembers*. Houghton-Mifflin, Boston. [page 161]

Schoener, T. W. 1982. The controversy over interspecific competition. *American Scientist* 70: 586–595. [page 100]

Shipley, A. E., and E. W. MacBride. 1901. *Zoology: An Elementary Textbook*. Macmillan, London. [page 83]

Treherne, J. 1983. *The Galapagos Affair*. Jonathan Cape, London. [page 260]

Wake, D. B. 2010. A festival for Rosemary and Peter Grant. In P. R. Grant and B. R. Grant (eds.), *In Search of the Causes of Evolution: From Field Observations to Mechanisms*, pp. 360–65. Princeton University Press, Princeton, NJ. [page 234]

Warwick, A. R. 1972. *The Phoenix Suburb: A South London Social History*. Blue Boar Press, Richmond, Surrey, UK. [page 4]

Webster, M. S. 2015. The messy processes of evolution through a 40-year lens. *Evolution* 69: 2246–47. [page 254]

Weiner, J. 1994. *The Beak of the Finch*. Alfred Knopf, New York. [pages 116, 151, 157, 292]

Wicks, Ben. 1988. *No Time to Wave Goodbye*. Stoddart, Toronto, Canada. [pages 7, 12]

Wiggins, I. L., and D. M. Porter. 1971. *Flora of Galápagos*. Stanford University Press, Stanford, CA. [page 107]

Wilson, E. O. 1994. *Naturalist*. Island Press, New York. [pages 59, 135]

Woram, J. M. 1989. Galápagos Island Names. *Noticias de Galápagos* 48: 22–32.

Zinsser, W. 2004. *Writing about Your Life: A Journey into the Past*. Da Capo Press, New York.

SUBJECT INDEX

BIRD INDEX